T0292274

Studies in Computational Intelligence

Volume 617

Series editor

Janusz Kacprzyk, Polish Academy of Sciences, Warsaw, Poland
e-mail: kacprzyk@ibspan.waw.pl

About this Series

The series "Studies in Computational Intelligence" (SCI) publishes new developments and advances in the various areas of computational intelligence—quickly and with a high quality. The intent is to cover the theory, applications, and design methods of computational intelligence, as embedded in the fields of engineering, computer science, physics and life sciences, as well as the methodologies behind them. The series contains monographs, lecture notes and edited volumes in computational intelligence spanning the areas of neural networks, connectionist systems, genetic algorithms, evolutionary computation, artificial intelligence, cellular automata, self-organizing systems, soft computing, fuzzy systems, and hybrid intelligent systems. Of particular value to both the contributors and the readership are the short publication timeframe and the worldwide distribution, which enable both wide and rapid dissemination of research output.

More information about this series at http://www.springer.com/series/7092

Witold Pedrycz · Giancarlo Succi
Alberto Sillitti
Editors

Computational Intelligence and Quantitative Software Engineering

 Springer

Editors
Witold Pedrycz
Department of Electrical and Computer
  Engineering
University of Alberta
Edmonton, AL
Canada

Alberto Sillitti
Center for Applied Software Engineering
Genova
Italy

Giancarlo Succi
Innopolis University
Innopolis
Russia

ISSN 1860-949X   ISSN 1860-9503   (electronic)
Studies in Computational Intelligence
ISBN 978-3-319-25962-8   ISBN 978-3-319-25964-2   (eBook)
DOI 10.1007/978-3-319-25964-2

Library of Congress Control Number: 2015957101

© Springer International Publishing Switzerland 2016
This work is subject to copyright. All rights are reserved by the Publisher, whether the whole or part
of the material is concerned, specifically the rights of translation, reprinting, reuse of illustrations,
recitation, broadcasting, reproduction on microfilms or in any other physical way, and transmission
or information storage and retrieval, electronic adaptation, computer software, or by similar or dissimilar
methodology now known or hereafter developed.
The use of general descriptive names, registered names, trademarks, service marks, etc. in this
publication does not imply, even in the absence of a specific statement, that such names are exempt from
the relevant protective laws and regulations and therefore free for general use.
The publisher, the authors and the editors are safe to assume that the advice and information in this
book are believed to be true and accurate at the date of publication. Neither the publisher nor the
authors or the editors give a warranty, express or implied, with respect to the material contained herein or
for any errors or omissions that may have been made.

Printed on acid-free paper

This Springer imprint is published by SpringerNature
The registered company is Springer International Publishing AG Switzerland

Preface

Undoubtedly, Software Engineering is an intensive knowledge-based endeavor of inherent human-centric nature, which profoundly relies on acquiring semiformal knowledge and then processing it to produce a running system. The knowledge spans a wide variety of artifacts, from requirements, captured in the interaction with humans, the customers, to design practices, testing, and code management strategies, which rely on the knowledge of the running system.

It is worth noticing that all the knowledge in place can be only partially formalized, and this relates not only to the interactions with customers to gather requirements, where the human factors make the impossibility of a complete formalization very evident, but also dealing with design and coding, where the intrinsic complexity of the underlying system, its evolvability and instability make unfeasible a precise and unique description of its features.

The chapters of this volume reflect upon a comprehensive body of knowledge positioned at the junction of the broadly perceived discipline of Software Engineering and Computational Intelligence. In particular, the visible pillars of the discipline embrace the following:

- Quantitative Software Engineering, which aims at building sound and practically viable methodologies of developing software artifacts,
- Nature of software models as nonlinear and highly adaptive architectures with a visible reliance on granular character of available data and knowledge,
- Global optimization commonly supported by evolutionary and population-based optimization technologies, delivering a necessary framework of structural and parametric optimization.

It is instructive to walk over the contributions of this volume. The two concise opening chapters cover the principles of Software Engineering and the associated relationship with Computational Intelligence (Chapter "The Role of Computational Intelligence in Quantitative Software Engineering") and fundamentals of Computational Intelligence (Chapter "Computational Intelligence: An Introduction"). They serve two main objectives. First, they set up a stage for more detailed and focused discussions presented in the sequence of chapters. Second, these chapters make the volume self-contained and help the reader acquire some useful prerequisites and establish a general conceptual and algorithmic setting.

Feature selection is an essential phase of any pattern classification scheme being crucial to the success of ensuing classifiers. As a matter of fact, selecting the best suite of features is of paramount relevance to the improved classification rate, prediction accuracy, and reduced computing overhead. There have been numerous approaches addressing this problem including a series of representative examples such as information gain attribute ranking (IG), Relief (RLF), principal component analysis (PCA), correlation-based feature selection (CFS), consistency-based subset evaluation (CNS), wrapper subset evaluation (WRP), and an evolutionary computation method including genetic programming (GP). Classification of artifacts of Software Engineering is not an exception to this rule and speaks loudly to the relevance of processes of feature selection. The chapter authored by Afzal and Torkar studies this important topic in the setting of software fault prediction models. The authors demonstrated quantifiable advantages of feature selection clearly manifesting in the increased classification accuracy.

Software testing is the critical development phase irrespectively of the specific methodology of software architecture and design and becomes of paramount relevance to the quality of the constructed software. In virtue of its nature, software testing is a highly computationally intensive process. Exhaustive testing is not feasible at all; hence, there is an ongoing quest to develop an efficient heuristics that reduces computing overheard while still retaining the effectiveness of a suite of tests. The study by Lopez-Heron, Fervier, Chiclano, Wegedyd, and Alba tackles the essence of this evident and timely challenge: The authors investigate a highly promising avenue of evolutionary computing as a vehicle supporting testing of software product lines. Along with a concise exposition of the key benefits of evolutionary computing, the authors identify key challenges associated with the specificity of software testing and the immensely overwhelming curse of dimensionality.

The study authored by Patrick delivers an authoritative treatment of the subject of software testing by focusing on test data generation where such data are generated via meta-heuristics and mechanisms of mutation. Mutation is a well-known testing approach that helps assess the abilities (quality) of existing suites of test cases to identify faults. It produces a large number of mutants—programs exhibiting some small syntactic change of the code. The key challenge is to form a suitable fitness function which helps assess the quality of this suite of tests. The chapter brings forward a large collection of fitness functions developed by nature-inspired algorithms.

The realm of Computational Intelligence, especially pattern recognition with its plethora of classification schemes, is investigated by Pizzi in his contribution devoted to the quantifying the utility of functional-based software. The quality of software components developed in this way is described by a series of software metrics and their labeling for classification purposes is completed with the use of centroid-based method and the technology of fuzzy sets.

In the chapter authored by Twala and Verner, a central problem of software cost estimation is analyzed with the use of multiple classifier systems. The paradigm of multiple classifiers has been shown to exhibit tangible benefits in numerous applications: Simple classifiers of relatively weak performance combined together (though bagging or boosting) tend to offer a significant improvement in terms of their combined classification performance. The study exemplifies the performance when using 10 industrial datasets and showing the improved performance in terms of the achieved smoothed error rate.

Tamir, Mueller, and Kandel study the technology of fuzzy sets in application to Quantitative Software Engineering Requirements specification. Requirements specification is very much a human-oriented process that helps eliciting, documenting, and formalizing software requirements about the tasks that the software must support. Generalizing this notion of viewing software from an operator's perspective yields the concepts of viewing software requirements from the external tasks. The authors propose a novel idea of complex fuzzy logic reasoning to articulate and prudently quantify intricate relations existing among various software engineering constraints including quality, software features, and development effort. Complex fuzzy logic has been around for some time and the authors demonstrate its abilities to capture and quantify the essence of expressing requirements.

The theory of possibility and necessity, being an integral part of the discipline of fuzzy sets, has been creatively investigated by Bellandi, Cimato, Damiani, and Gianini. The thorough discussion convincingly articulates the need of granular constructs (fuzzy sets) along with the possibilities and necessities to quantify and characterize risk assessment. The detailed studies are focused on disclosure risk assessment in cloud processes described by a possibilistic model of information disclosure. The ensuing case study concerns cloud-based computations.

We really enjoyed working on this project on a timely, highly relevant and truly multidisciplinary and hope the readers will find this volume informative and helpful in pursuing research and applied studies in a creative exploration of the concepts, methodology, and algorithms of Computational Intelligence in Software Engineering.

Our thanks go to the contributors of this volume whose expertise and enthusiasm made this project successful. We would like to express our thanks to Professor Janusz Kacprzyk, Editor-in-Chief of this book series for his continuous support, vision, and to bring advanced technologies of Computational Intelligence and their recent and the most successful practices. It was our genuine pleasure working with the professional staff of Springer throughout the realization of this publication project.

Support from the Department for Educational Policies, Universities, and Research of the Autonomous Province of Bolzano, South Tyrol, to Witold Pedrycz is gratefully acknowledged.

June 2015   Witold Pedrycz
                                                                        Giancarlo Succi
                                                                        Alberto Sillitti

Contents

The Role of Computational Intelligence in Quantitative Software Engineering

Witold Pedrycz, Alberto Sillitti and Giancarlo Succi

Software development has been often considered as a "standard" manufacturing activity, whose actions can be sequenced and optimized quite like the production of cars. From this the "Waterfall Model" of software production was defined. But, like most human activities, even what people consider a "simple" production of a Cappuccino, cannot be represented as such, and software is definitely more difficult than making a Cappuccino; in particular, in software three major problems occur: irreversibility, uncertainty, and complexity. This has lead to the emergence of novel approach to software production that acknowledges such problems and reverts upside-down the idea of formalizing software processes with a waterfall approach. However, such acknowledgment has not yet been fully received in the area of empirical and quantitative software engineering, where software processes have been still modeled usually with standard statistics and other traditional mathematical tools. We advocate that the usage of Computational Intelligence can translate into empirical and quantitative software engineering the paradigm shift implemented by the new, emerging process models in software development.

W. Pedrycz (✉)
Department of Electrical and Computer Engineering, University of Alberta,
Edmonton, AB T6R 2V4, Canada
e-mail: wpedrycz@ualberta.ca

W. Pedrycz
Department of Electrical and Computer Engineering, Faculty of Engineering,
King Abdulaziz University, Jeddah 21589, Saudi Arabia

A. Sillitti
Center for Applied Software Engineering, Genova, Italy

G. Succi
Innopolis University, Innopolis, Russia

© Springer International Publishing Switzerland 2016
W. Pedrycz et al. (eds.), *Computational Intelligence and Quantitative Software Engineering*, Studies in Computational Intelligence 617,
DOI 10.1007/978-3-319-25964-2_1

1

1 Introduction—Software Development and the Art of Cappuccino

Software development is not a simple and straightforward process; it must be approached in an organized way. We illustrate some of the organizational issues with an extended example—preparing a cup of cappuccino.

Suppose you want to have a good cup of cappuccino and suppose you are at home, so you cannot go to Starbucks and get one! Preparing a cappuccino requires a sequence of operations. We can represent such operations with a flowchart. After all, we are software engineers, so flowcharts are quite familiar to us.

The graph in Fig. 1 makes a lot of sense. It is clear, and unambiguous. When we follow its well-defined set of operations, we accomplish our goal of having an excellent cup of cappuccino, or … we think (hope?) we do…

However, there is an additional fact that doesn't appear in the chart: we should not let the espresso get cold while we steam the milk.

Indeed, describing how to draw this step is not a problem for us. After all, we are software engineers and we know how to draw a Petri net (Fig. 2).

The horizontal line represent points at which the activities are synchronized. The execution of the tasks proceeds only when *all* the activities above the bar ("entering" the bar) have been completed.

Well, now we are ready to get our MSC.[1]

However, if you try to accomplish the cappuccino-making tasks as shown in Fig. 2, you will realize that there are a few problems still to cope with:

1. If you live in the mountains, it may be possible that your cup of espresso is not as good as that the cup you might enjoy at the sea level: The water boils earlier at higher elevations, so you might lose some of the aroma.
2. The choice of the coffee from which you make the espresso has a significant impact on the resulting quality of the cappuccino; the same for the quality of the water you use, (sometimes bottled water is required) and the same for the kind of milk.
3. The basic resources are perishable. Once you have made a bad espresso, there is no way to undo the process and get back the original coffee. The coffee is simply wasted. Moreover, the milk cannot be steamed too many times. After a while, steaming the milk does not produce any more foam; the milk is just gone.
4. The magic touch! You need a special touch to make a good cappuccino. The touch is needed when you prepare the espresso. Otherwise the espresso would become a Lungo or an Americano. The touch is needed when you have the foam ready and you pour it in the cup. Without such touch, the cappuccino could well become just a Latte.

---

[1]Master in the Science of Cappuccino.

Fig. 1 A chart explaining how to prepare a cup of cappuccino

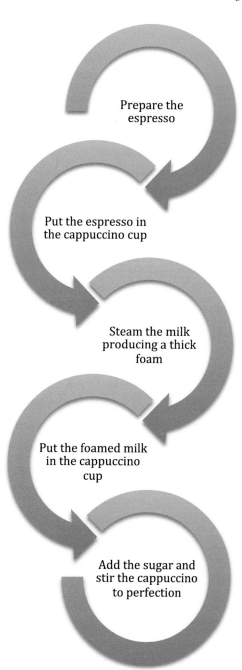

Prepare the espresso

Put the espresso in the cappuccino cup

Steam the milk producing a thick foam

Put the foamed milk in the cappuccino cup

Add the sugar and stir the cappuccino to perfection

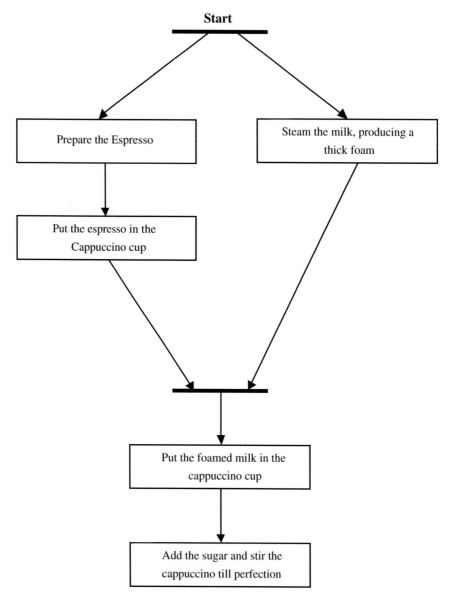

Fig. 2 A Petri net explaining how to prepare a cup of cappuccino

To make evident that the inability of formalizing properly the production process is not just a problem of a few coffee-enthusiasts, we can also consider the product lifestyle used by civil engineers who build homes. In building a home, the engineer and the future homeowners follow a few essential steps, which were learned and refined from the experience of several centuries of building and owning houses. Such steps can be summarized as follows.

- Analyze the requirements of the customers.
- Design the home.
- Purchase the land on which to build the home.
- Commission the development to a builder.
- Check the resulting building for conformance to our requirements and to the municipal, provincial, and federal standards.
- Maintain the house through regular cleaning, restorations, and additions.

Notice that also in this case:

- The lifecycle is based on experience and does not depend on some predefined model.
- The lifecycle stages aim at reducing uncertainty in the use of the product and at reducing the complexity in the overall development.
- The stages are changeable; they could be altered if better or more suitable stages could be found.

2 Persistent Problems in Software Development

Altogether, we can claim that both in the case of the cappuccino and in the case of the construction of buildings, not all factors are under our control. We cannot travel to get to sea level for a better cup of cappuccino. Worse, not all of the factors are predictable. Look at the magic touch or at the ever-changing desires of the customers. You can identify its presence or absence only after the cappuccino making is complete or the home is fully built. In fact, the entire process of preparing a good cappuccino is permeated by a terrible characteristic, one that makes us very nervous each time we hear it: "uncertainty." This is why, once we do realize to have found an excellent cappuccino shop, we stick to it with a religious faith. And this is practically the same when we build a home.

Moreover, there is another issue to consider: "We always get older!" There is no way to revert this irreversible passage of time, despite years of medical studies, and the efforts of Superman, Wonder Woman, and Austin Powers. And there is no way to rescue a bad espresso: the coffee is gone. And even worse, after a while, there is no way to produce foam from already steamed milk! So, overall there are situations that are irreversible throughout the production of cappuccino. Likewise, once we have build a home, we cannot move it by 50 cm more north or rotate it a bit more so that the master bedroom looks properly at the rising sun.

Now, suppose you get good in producing espressos and you start building a little business for your classmates. You are now faced with the problem of preparing ten cappuccinos at the same time. You might be tempted to ask for help. You might even think to ask your boyfriend or girlfriend to assist you. Do not do it! If you did,

Table 1 Examples of uncertainty, irreversibility, and coordination throughout software development

	Before development	During development	After development
Uncertainty in	The desires of customer; the business goals of managers; the personal goals of developers	The resources required to develop the system; the reliability of the tools and the libraries used for the system	The failures of the system; the possibility for the system to work in different OSs and architectures
Irreversibility in	The definition of the expected level of reliability	The architecture of the system; structures of packages	The level of support decided; upgrading to a new platform
Complexity in	Customers, managers, marketing people, and analysts	Cowboy coding, spaghetti code	Customers, customer service, developers, and managers

you would realize soon that you would have to decide upfront the tasks for each one of you, determine when you need to accomplish such tasks, and how you can synchronize your different actions. This requires time and discipline. If you do not believe us, go to the closest Starbucks and watch at what the waiters do! We call this the problem of "complexity."

Successful software managers, engineers, and developers need to cope with these three factors to deliver software "on time and within budget" [8]. Uncertainty, irreversibility and coordination are present throughout the life of software systems, before, during, and after development. Table 1 summarizes these relationships, and each is further explained in the sections that follow.

3 Uncertainty

Before development, uncertainty comes from the difficulty of communication among customers, among developers, and between customers and developers. Uncertainty is present before development in the expectations of the customers: often the customer has not set his/her mind to what is really wanted; often the customer does not manage to express what s/he wants in terms understandable by the developers. Or, the business goals of the managers and the personal goals of developers may be not entirely clear to everyone in the team. Such goals may not be aligned and thus not point to customer satisfaction.

During development, the developers may have the unpleasant surprise of finding that the supporting tools do not work as they expected, that the libraries do not adhere to the available documentation. The resources required to develop the system may not be available or reliable.

After development, the user may find that the system crashes every now and then, with a quite unpredictable pattern. In addition, porting the system to a new PC or upgrading the operating system may result in the system not working as expected.

4 Irreversibility

Irreversibility comes from the impossibility of reverting to a previous state. In a sense, we always experience irreversibility: there is no way to reverse time! A major source of irreversibility lies in the fact that resources spent never come back. So if we waste useless time in doing work that later appear to be not needed, such time is not recoverable easily. This aspect of irreversibility is closely related to the problem of uncertainty. We do not know upfront all what will be needed. So we estimate, and our ability to estimate well reduces the waste. However, estimates may depend on factors beyond our knowledge.

Before development, irreversible decisions may be made when the customer specifies a given quality of the resulting system. For example, the customer may consider acceptable a tool that crashes every second hour. Later, it may be difficult to increase the expected quality of the software system. An even worse situation arises when the customer does not specify such a quality level, but developers incorrectly make assumptions of what is acceptable.

During development, irreversible decisions can permeate the entire process, from deciding the architecture of the system, to the selection of design patterns, down to the structure of the unit tests. Each one of these decisions commits the developer to subsequent choices and blocks other choices. Undoing such decisions is expensive, even more with passing time.

After development, irreversible decisions can also include a decision to upgrade or not to upgrade a software system to a new architecture, this may introduce irreversible mechanisms that affect the future life of the system.

5 Complexity

Complexity comes from the Latin words "cum," meaning "with," and "plexo," meaning "put together." It expresses the difficulty of mastering all the details of a very intricate issue.

In software projects, the number of decisions that must be is much higher than in any other field. After all, we call the product "soft"ware to evidence that we deal with soft entities, that are easy to modify and rearrange according to our needs. There is therefore a high risk of ending with an intractable complexity.

Before development, complexity is high. Coordinating different—any persons who are interested in the system—is a very complex task involving customers (who pay for the system), users (who use the system), managers, designers, developers, marketing people, and so on. Successful development requires aligning the goals of these multitudes of people to the common end of a working system in time and on budget.

After development, uncertainty may persist. Uncertainty and irreversibility make it hard to synchronize plans completely early in the lifecycle of a project. Therefore, such coordination among stakeholders is required throughout the development cycle, before, during, and after.

6 Handling Uncertainty, Irreversibility, and Complexity ... and Cappuccino!

Indeed, there have been multiple proposals to handle Uncertainty, Irreversibility, and Complexity in Software Development. Among the most notable, there have been the efforts of the whole agile movement. To get to that point, we revisit now the example of the preparation of the Cappuccino.

The product lifecycle of a cup of cappuccino at a good cappuccino shop is the following:

1. Training of waiters: The waiters are taught how to prepare the cappuccino.
2. Acquisition of raw materials: Fresh milk and Arabic coffee beans are purchased.
3. Preparation of the ingredients: The espresso is prepared and the milk is foamed.
4. Finalization of the cappuccino: The espresso is put in the cup followed by the foamed milk.
5. Delivery: The cup of cappuccino is served to the customer.
6. Consumption: The customer enjoys the cappuccino.

A good product relies on the understanding and appreciation of all the people involved in its production and in its consumption.

- Cappuccino developers understand "the big picture" of the preparation of a cup of cappuccino and learn how to make it.
- Cappuccino consumers appreciate the complex process required to develop a cappuccino and are patient for the few extra minutes required to prepare a good cappuccino, rather than quickly grabbing a regular American coffee that can be served instantaneously.
- A clear difference can be drawn between good cappuccino shops and standard cappuccino shops (the one where an instantaneous pseudo-cappuccino comes from a strange, metallic box).

Good cappuccino shops try to explain such steps to their customers, making the process of preparing a cappuccino a ceremony visible to everyone, and providing leaflets on coffee, espresso machines, and related materials.

Formalizing the steps in the process is more or less what the ISO 9000 quality process requires. Please refer to the later chapter of this book for a more detailed description.

Altogether, the definition of the lifecycle for a product reduces the effects of the three beasts of uncertainty, irreversibility, and complexity:

- It reduces the uncertainty of customers, because descriptions are provided stating what will be done.
- It reduces the complexity of development, because descriptions are provided stating what will be done and why.

Also, the lifecycle stages for a product may not always be the same in every context. In the case of the cup of cappuccino, you go to one of the most fashionable coffee shops in Rome, the "Caffè Sant'Eustachio," you will see a different process for cappuccino. The actual process is kept secret and the cappuccino is developed in the dark of a little closet. The stages are just steps learned from experience. That experience ensures the consistency of the final results.

Starting with the late '90 people have started to target this problem with alternative models that appear to stem form the experiences of Lean Management [12] in manufacturing and of Spiral Development Models in Software Engineering [4]: the so-called Agile Models [1], which have emerged and acquired a prominent role. Agile Models of software development, like XP [3], Scrum [9], Kanban [2] are now very common in the industry [10] and are part of most software engineering curricula. They are based on the idea of eliminating the "waste" from software production, including the expensive long-term planning sessions and large upfront design activities, to focus on what is predictable with reliability and developable with a reliable understanding that will be effectively useful in the construction of what is desired by the customer. Moreover, Agile Methods focus on establishing a high band of communication with the customer, so that a better mutual understanding of what is desired (from the side of the customer) and what is feasible (from the side of the developer) is achieved, with the positive resulting implications of the system under development.

7 The Pivotal Role of Computational Intelligence in Quantitative Software Engineering

So far we have seen that the traditional "divide-and-conquer" approach of the so-called scientific management falls short in the preparation of the cappuccino and in the construction of homes and does not hold for software development. This concept is now largely diffused and accepted.

Still, when we deal with modeling and interpreting software development processes, when we try to determine the effort required to build a software system or its resulting quality, we still rely on pretty conventional mathematical abstractions, which do not take into account the presence of complexity, irreversibility, and uncertainty.

This is the place where Computational Intelligence plays a pivotal role, since:

- It expresses **uncertainty** by its own nature. Just as an example, fuzzy systems represent well elements whose natures are not defined in a two-valued Boolean way. Consider the case of requirements: only at the end of the project we know which requirements have been implemented, and only partially, as we know from the halting problem. Still, at the beginning of a project, regardless of the lifecycle model, requirements are assigned priority, planned resources, etc. Needless to say, all such *numerifications* of the problem are by their own nature mistake, since they do not take into account the fact that the reliability of such number is often similar to the one of an unprofessional psychic[2] predicting the future.
- It handles **complexity** by its own nature. Just as an example, neural networks are able to model the behaviours of systems by observing them and learning from such observations. Therefore, neural network are suitable when a formal and explicit modeling is not derivable. This is the case, for instance, in building predictive models of the effort and the time to produce a software system: in this area there have been countless proposals so far, but the only comprehensive models date decades and, indeed, are absolutely unusable in practical contexts.
- It incorporates the nature of **irreversibility**, that is, the fact that solutions are typically non-linear and, therefore, not associable and dissociable, i.e., that the path leading to a solution influences also the solution which is found. This is what happens in typical problems of minimization in computational intelligence, where the initial conditions indeed impact the final solution. A clear example of this in the production of software systems is the selection of the order of implementation of requirements. There are researchers who have assumed a hedonic model, where we assign a cost to each requirement and then whatever is the sequence of their implementation, then the final cost is the sum of the partial costs. This is simply false. And, please note that we are not referring to the clear case when, for instance, to build a transaction on a database we need to have the database in place—in this case it is obvious that building the transaction before the database would result in much higher costs. We are rather referring to the situation of allegedly independent requirements: implementing a requirement and, if the case, showing it to the customer, has the potential and the likelihood to alter the developers and, if it is the case, the customers perceptions of the system being built, and thus, how the future requirements will be taken care of.

People might be puzzled why, if our argumentation is true, computational intelligence has not been implemented more often and systematically in Software Engineering. Our idea is that it is a learning process that follows with some delay the fate of the various proposal for software lifecycles to the proposals of how to quantify aspects of the production process building suitable models. Therefore, after

---

[2]Professional psychic in general are better in this, since they are able to express their predictions in a very dubitative form, which makes them useless but rarely falsifiable.

The Role of Computational Intelligence …   11

about a decade and a half from the beginning of the heavy diffusion of Agile Methods in the industry, it is now the time that Computational Intelligence takes the leading role, offering to researchers the suitable models and techniques to build quantitative models of software production, this achieving higher reliability in process control and estimation, which would result in better selecting also the lifecycle models to adopt in the various situations; to this end some effort has already started [6].

8 Conclusions

The production of even of simple artifact, like a cappuccino, can be hardly represented by formal mathematical elements and managed accordingly. Software development, which is far from being simple, repels from any crisp representation and management. To this end, innovative methods of handling software production have been proposed, like Agile Methods.

If it is not possible to manage the software in such "scientific way," it is completely unfeasible the idea to represent software with linear models and/or well defined formulas. This acknowledgement paves the way for the adoption of techniques from Computational Intelligence, which model more effectively and reliably unclear and uncertain situations, like the ones typical of software.

While there have been proposals already in the past, we think that only now is the proper time to run deep and systematic investigations in the use of these techniques in software, since:

- there has been a full acknowledgement in the technical and scientific community of the unfeasibility of scientific management in software,
- after years of scarce availability of software engineering data, also thanks to the adoption of non invasive software measurement techniques, there is finally an abundance of it, that makes possible to perform analysis previously unfeasible.

The hope is that with the adoption of suitable techniques of Computational Intelligence [7], it will be finally possible to answer the typical questions of software production, like whether Pair Programming is indeed beneficial [5, 10], in which context it is suitable to perform upfront design and when, in converse, it is better to perform pure incremental development [11].

References

1. Agile Manifesto: Manifesto for Agile software development (1999). URL: http://agilemanifesto.org. Visited on the 21 May 2015
2. Anderson, D.J.: Kanban: Successful Evolutionary Change for Your Technology Business. Blue Hole Press, USA (2010)

3. Beck, K.: Extreme Programming Explained: Embrace Change. Addison Wesley, Reading (1999)
4. Boehm, B.W.: A spiral model of software development and enhancement. IEEE Comput. **21** (5), 61–72 (1988)
5. Coman, I.D., Sillitti, A., Succi, G.: Investigating the usefulness of pair-programming in a mature agile team. In: Agile Processes in Software Engineering and Extreme Programming, Proceedings of XP2008, pp. 127–136. Springer, Berlin
6. Fronza, I., Sillitti, A., Succi, G., Terho, M., Vlasenko, J.: Failure prediction based on log files using random indexing and support vector machines. J. Syst. Softw. **86**(1), 2–11 (2013)
7. Pedrycz, W., Succi, G., Sillitti, A., Iljazi, J.: Data description: a general framework of information granules. Knowl. Based Syst. **80**, 98–108 (2015)
8. Putnam, L.H., Myers, W.: Measures for Excellence: Reliable Software on Time. Within Budget, Yourdon (1992)
9. Schwaber, K.: Agile Project Management with Scrum. Microsoft Press, USA (2004)
10. Sillitti, A., Succi, G., Vlasenko, J.: Understanding the impact of pair programming on developers attention: a case study on a large industrial experimentation. In: Proceedings of the 34th International Conference on Software Engineering, Zurich, CH, pp. 1094–1101
11. Valerio, A., Succi, G., Fenaroli, M.: Domain analysis and framework-based software development. Appl. Comput. Rev. **5**(2), 4–15 (1997)
12. Womack, J.P., Jones, D.T.: Lean Thinking: Banish Waste and Create Wealth in Your Corporation. Productivity Press, Revised and Updated (2003)

Computational Intelligence:
An Introduction

Witold Pedrycz, Alberto Sillitti and Giancarlo Succi

Abstract The study offers an introduction to the paradigm, concepts and algorithms of Computational Intelligence (CI). We elaborate on the main technologies of CI: neural networks, fuzzy sets or Granular Computing, in general, and evolutionary optimization and identify their focal points and stress an overall synergistic character of these technologies, which ultimately gives rise to the highly symbiotic processing environment. Furthermore, the main advantages and limitations of the CI technologies are discussed. The key linkages of CI with the area of Software Engineering, especially its quantitative facet, are stressed.

Keywords Computational intelligence · Neurocomputing · Fuzzy sets · Information granules · Granular computing · Interpretation · Synergy · Software engineering

1 Introduction

In this study, we discuss the main conceptual, methodological and algorithmic pillars of Computational Intelligence (CI), identify their main features and elaborate on their role in biomedical signal processing. To a significant extent, the content of

W. Pedrycz (✉)
Department of Electrical and Computer Engineering, University of Alberta,
Edmonton, AB T6R 2V4, Canada
e-mail: wpedrycz@ualberta.ca

W. Pedrycz
Department of Electrical and Computer Engineering, Faculty of Engineering,
King Abdulaziz University, Jeddah 21589, Saudi Arabia

A. Sillitti
Center for Applied Software Engineering, Genova, Italy

G. Succi
Innopolis University, Innopolis, Russia

© Springer International Publishing Switzerland 2016   13
W. Pedrycz et al. (eds.), *Computational Intelligence and Quantitative
Software Engineering*, Studies in Computational Intelligence 617,
DOI 10.1007/978-3-319-25964-2_2

this study is self-contained and the most essential ideas are elaborated on from scratch. The reader can benefit from some introductory knowledge of the subject matter on neural networks, fuzzy sets and evolutionary computing; see, for instance [25, 26]. The presentation is structured in a top-down manner. We start with a concise introduction to Computational Intelligence (CI) being viewed as a highly synergistic environment bringing a number of highly visible technologies of Granular Computing, neural networks, and evolutionary optimization (Sect. 2). In a sequence of sections, Sects. 3–5, we discuss neurocomputing, evolutionary optimization, and Granular Computing. Furthermore we show that the three main technologies of CI are naturally inclined to foster and exploit useful synergistic linkages. Formal platforms of information granularity are discussed in Sect. 6. We elaborate on information granularity and its role in signal representation in Sect. 7. The concept of information granulation-degranulation is discussed in Sect. 8. The design of information granules regarded as semantically sound abstractions is covered in Sect. 8. Here we discuss ways where not only numeric data—experimental evidence is taken into account but various tidbits of domain knowledge are also used in the formation of information granules. In Sect. 9, we visualize a role of CI in software engineering.

With regard to the notation used in this study is concerned we follow the symbols being in common usage. Patterns (data) $\mathbf{x}_1, \mathbf{x}_2, \ldots, \mathbf{x}_N$ are treated as vectors located in n-dimensional space $\mathbf{R}^n$, $\|.\|$ is used to denote a distance (Euclidean, Mahalanobis, Hamming, Tchebyshev, etc.). Fuzzy sets will be described by capital letters; the same notation is being used for their membership functions.

2 Computational Intelligence: An Agenda of Synergy of Algorithms of Learning, Optimization and Knowledge Representation

Computational Intelligence can be defined in many different ways. Let us start by recalling two definitions or descriptions, which are commonly encountered in the literature

A system is computationally intelligent when it: deals with only numerical (low-level) data, has pattern recognition components, does not use knowledge in the AI sense; and additionally when it (begins to) exhibit (1) computational adaptivity; (2) computational fault tolerance, (3) speed approaching human-like turnaround, and (4) error rates that approximate human performance [5, 6]

The description provided by W. Karplus comes as follows

CI substitutes intensive computation for insight how the system works. Neural networks, fuzzy systems and evolutionary Computation were all shunned by classical system and control theorists. CI umbrellas and unifies these and other revolutionary methods

The first description captures the essence of the area. Perhaps today such a definition becomes slightly extended by allowing for some new trends and technologies, which are visible in the design of intelligent systems. Nevertheless the essence of CI is well-captured.

The comprehensive monograph on CI [26] emphasizes the importance of synergy of the contributing and very much complementary technologies of fuzzy sets, neurocomputing and evolutionary optimization. In a nutshell, CI is about effective and omnipresent mechanisms of synergy exploited in a variety of tasks of analysis and design of intelligent systems. The reader may refer to Fulcher and Jain [8] and Mumford and Jain [15], which serve as comprehensive sources of updated material on Computational Intelligence.

The emergence of CI is justifiable, and in some sense, unavoidable. Over time, being faced with more advanced problems, increased dimensionality and complexity of systems one has to deal with, neural networks, fuzzy sets and evolutionary computing started to exhibit some clear limitations. This is not startling at all as their research agendas are very much distinct and they focus on different aspects of the design of intelligent systems. The synergistic environment, in which knowledge representation, learning and global optimization go hand in hand, becomes a highly justifiable development environment.

Let us discuss in more detail on knowledge representation as being captured by fuzzy sets. Fuzzy sets offer a unique and general opportunity to look at information granules as semantically meaningful entities endowed with detailed numeric description. For instance, consider an information granule termed *high* amplitude of signal. On the one hand, *high* is just a single symbol and as such could be processed at the level of symbolic processing encountered in Artificial Intelligence (AI). For instance, it could be one of the symbols used in syntactic pattern classifier captured by a collection of syntactic production rules or automata. On the other hand, the same granular entity *high* is associated with the detailed numeric description, which calibrates the concept in presence of available numeric evidence. A suitable level of abstraction helps us establish the most promising tradeoff between detailed and abstract view at the problem/data. Of course, the choice of the tradeoff is problem driven and depends upon the task and the main objectives specified therein. Likewise, the same information granule *high* can be looked at in less detail and through the use of some partially specified numeric content (that is in the form of higher type information granules, say fuzzy sets of type-2) could be processed in a semi-numeric fashion. In this way, the granularity of information and a formal mechanism used in granulation itself offers a way to position anywhere in-between symbolic view and numeric perception or quantification of the reality.

One may emphasize an important and enlightening linkage between Computational Intelligence and Artificial Intelligence (AI). To a significant extent, AI is a synonym of symbol-driven processing faculty. CI effectively exploits numeric data however owing to the technology of Granular Computing, it may invoke computing based on information described at various levels of granularity by inherently associating such granules with their underlying semantics described in a numeric or semi-numeric fashion (such as e.g., membership functions, characteristic functions or interval-valued mappings). The granularity of results supports the user-friendly nature of CI models. They can also form an important construct to be further used in facilitating interaction with the user as well as forming linkages with symbolic processing of AI constructs. The three fundamental components of CI along with an emphasis on their synergy of the main function give rise to a plethora of architectures in which various technologies assume a dominant role. This comes with various names such as neurofuzzy systems, evolutionary neural networks, genetic neural classifiers, etc. Let us recall that knowledge representation associated with the component of information granularity (along with its abstraction facet), learning (adaptive) abilities and global structural optimization (supported by evolutionary methods).

3   Neural Networks and Neurocomputing

There exists an immensely vast body of literature on neural networks. Neural networks are viewed as highly versatile distributed architectures realizing a concept of universal approximation [10, 27], which offers a very much attractive feature of approximating nonlinear (continuous) mappings to any desired level of accuracy and in this way supporting various classification tasks.

The two main taxonomies encountered in neurocomputing can be established centered around: (a) topologies of networks and (b) a variety of ways of their development (training) schemes. With regard to the first coordinate of the taxonomy, one looks at a way in which individual neurons are arranged together into successive layers and a way in which processing is realized by the network, namely if this is of feedforward nature or there are some feedback linkages within the structure. Typically, within the spectrum of learning scenarios one distinguishes between supervised learning and unsupervised learning however there are a number of interesting learning schemes, which fall in-between these two extreme positions (say, learning with partial supervision, proximity-based learning, etc.).

One needs to be aware of some limitations of neural networks that start manifesting in practical scenarios (those drawbacks might be alleviated to some extent but it is unlikely they will vanish completely). From the perspective of practice of neural networks in the context of biomedical signal processing, we can list a number of advantages. The main of those include: universal approximation capabilities (neural networks are universal approximators), significant learning abilities with a large repository of algorithms, well-developed and validated training methods. Neural networks support distributed processing, which as a result exhibit high potential for endowing them with significant fault tolerance capabilities. Just recently, there is a visible interest in the efficient realizations of networks, especially when considering their usage in portable medical devices. Along with these evident advantages, one has to be aware of several limitations of which neural networks are not free from. Neural networks exhibit black-box architectures (in other words, there is some effort to interpret constructed networks). Gradient-based optimization exhibits all limitations associated with this type of learning.

We witness non-repetitive results of learning of the networks (depending upon initial learning condition, parameters of the learning algorithm, etc.) while the learning realized in the presence of high-dimensional and large data sets could be slow and inefficient.

We should highlight that by no means neural networks can be sought as a plug-and-play technology. To the contrary: its successful usage does require careful planning, data organization and data preprocessing, a prudent validation and a careful accommodation of any prior domain knowledge being available. The black box nature of neural networks can bring some hesitation and reluctance to use the neural network solution and one has to be prepared for further critical evaluation of the obtained results.

4 Evolutionary and Biologically Inspired Computing: Towards a Holistic View at Global Optimization

The attractiveness of this paradigm of computing stems from the fact that all pursuits are realized by a population of individual potential solutions so that this offers a very much appealing opportunity of exploring or exploiting a search space in a holistic manner [9]. The search is realized by a population—a collection of individuals, which at each iteration (generation) carry out search on their own and then are subject to some processes of interaction/communication.

In case of genetic algorithms, evolutionary methods, and population-based methods (say, genetic algorithms, evolutionary strategies, particle swarm optimization), in general, a population undergoes evolution; the best individuals are retained, they form a new population through recombination. They are subject to mutation. Each operator present in the search process realizes some mechanism of exploration or exploitation of the search space. A general processing scheme can be outlined as follows

{evaluate population (individuals)
select mating individuals (selection process)
recombination
mutation}

The above basic sequence scheme is repeated (iterated).

In contrast to evolutionary methods, in the swarm-based methods [7], we encounter an interesting way of sharing experience. Each particle relies on its own experience accumulated so far but it is also affected by the cognitive component where one looks at the performance of other members of the population as well as an overall behavior of the population.

The essential phase of any evolutionary and population-based method (directly affecting its performance) is a representation problem. It is concerned about a way how to represent the problem in the language of the search strategy so that (a) the resulting search space is made compact enough (to make the search less time consuming) and (b) is well reflective of the properties of the fitness function to be optimized. By forming a suitable search space we pay attention to avoid forming extended regions of the search space where the fitness function does not change its values.

The *forte* of the methods falling under the rubric of these population-based optimization techniques is the genuine flexibility of the fitness function—there is a great deal of possibilities on how it can be formulated to capture the essence of the optimization problem. This translates into an ability to arrive at a suitable solution to the real-world task.

The inevitable challenges come with the need to assess how good the obtained solution really is and a formation of the feature space itself.

5   Information Granularity and Granular Computing

Information granules permeate numerous human endeavors [1, 3, 19, 29, 30]. No matter what problem is taken into consideration, we usually express it in a certain conceptual framework of basic entities, which we regard to be of relevance to the problem formulation and problem solving. This becomes a framework in which we formulate generic concepts adhering to some level of abstraction, carry out processing, and communicate the results to the external environment. Consider, for instance, image processing. In spite of the continuous progress in the area, a human being assumes a dominant and very much uncontested position when it comes to understanding and interpreting images. Surely, we do not focus our attention on individual pixels and process them as such but group them together into semantically meaningful constructs—familiar objects we deal with in everyday life. Such objects involve regions that consist of pixels or categories of pixels drawn together because of their proximity in the image, similar texture, color, etc. This remarkable and unchallenged ability of humans dwells on our effortless ability to construct information granules, manipulate them and arrive at sound conclusions. As another example, consider a collection of time series. From our perspective we can describe them in a semi-qualitative manner by pointing at specific regions of such signals. Specialists can effortlessly interpret various diagnostic signals including ECG recordings. They distinguish some segments of such signals and interpret their combinations. Experts can interpret temporal readings of sensors and assess the status of the monitored system. Again, in all these situations, the individual samples of the signals are not the focal point of the analysis and the ensuing signal interpretation. We always granulate all phenomena (no matter if they are originally discrete or analog in their nature). Time is another important variable that is subjected to granulation. We use seconds, minutes, days, months, and years. Depending upon a specific problem we have in mind and who the user is, the size of information granules (time intervals) could vary quite dramatically. To the high-level management time intervals of quarters of year or a few years could be meaningful temporal information granules on basis of which one develops any predictive model. For those in charge of everyday operation of a dispatching plant, minutes and hours could form a viable scale of time granulation. For the designer of high-speed integrated circuits and digital systems, the temporal information granules concern nanoseconds, microseconds, and perhaps microseconds. Even such commonly encountered and simple examples are convincing enough to lead us to ascertain that (a) information granules are the key components of knowledge representation and processing, (b) the level of granularity of information granules (their size, to be more descriptive) becomes crucial to the problem description and an overall strategy of problem solving, and (c) there is no universal level of granularity of information; the size of granules is problem-oriented and user dependent.

What has been said so far touched a qualitative aspect of the problem. The challenge is to develop a computing framework within which all these representation and processing endeavors could be formally realized. The common platform emerging within this context comes under the name of Granular Computing. In essence, it is an emerging paradigm of information processing. While we have already noticed a number of important conceptual and computational constructs built in the domain of system modeling, machine learning, image processing, pattern recognition, and data compression in which various abstractions (and ensuing information granules) came into existence, Granular Computing becomes innovative and intellectually proactive in several fundamental ways.

- It identifies the essential commonalities between the surprisingly diversified problems and technologies used there, which could be cast into a unified framework known as a granular world. This is a fully operational processing entity that interacts with the external world (that could be another granular or numeric world) by collecting necessary granular information and returning the outcomes of the granular computing
- With the emergence of the unified framework of granular processing, we get a better grasp as to the role of interaction between various formalisms and visualize a way in which they communicate.
- It brings together the existing formalisms of set theory (interval analysis) [14], fuzzy sets [28, 30], rough sets [16–18] under the same roof by clearly visualizing that in spite of their visibly distinct underpinnings (and ensuing processing), they exhibit some fundamental commonalities. In this sense, Granular Computing establishes a stimulating environment of synergy between the individual approaches.
- By building upon the commonalities of the existing formal approaches, Granular Computing helps build heterogeneous and multifaceted models of processing of information granules by clearly recognizing the orthogonal nature of some of the existing and well established frameworks (say, probability theory coming with its probability density functions and fuzzy sets with their membership functions)
- Granular Computing fully acknowledges a notion of variable granularity whose range could cover detailed numeric entities and very abstract and general

information granules. It looks at the aspects of compatibility of such information granules and ensuing communication mechanisms of the granular worlds.

• Interestingly, the inception of information granules is highly motivated. We do not form information granules without reason. Information granules arise as an evident realization of the fundamental paradigm of abstraction.

Granular Computing forms a unified conceptual and computing platform. Yet, it directly benefits from the already existing and well-established concepts of information granules formed in the setting of set theory, fuzzy sets, rough sets and others. While Granular Computing offers a unique ability to conveniently translate the problem into the language of information granules, it also exhibits some limitations associated with the lack of effective learning schemes, and quite commonly prescriptive nature of granular constructs (so there might be some danger of not carefully considering experimental evidence).

While all the three classes of technologies discussed so far offer tangible benefits and help address various central problems of intelligent systems, it becomes apparent that they are very much complementary. The strength of one technology is a quite visible limitation of some other. It is not surprising that there have been various ways of forming hybrid approaches dwelling upon the complementarity of neurocomputing, fuzzy sets (Granular Computing), and evolutionary methods, out of which a concept of Computational Intelligence (CI) has emerged.

6 Formal Platforms of Information Granularity

There exists a plethora of formal platforms in which information granules are defined and processed.

Sets (intervals) realize a concept of abstraction by introducing a notion of dichotomy: we admit element to belong to a given information granule or to be excluded from it. Sets are described by characteristic functions taking on values in $\{0,1\}$. A family of sets defined in a universe of discourse $\mathbf{X}$ is denoted by $P(\mathbf{X})$.

Fuzzy sets [28, 30] offer an important generalization of sets. By admitting partial membership to a given information granule we bring an important feature which makes the concept to be in rapport with reality. The description of fuzzy sets is realized in terms of membership functions taking on values in the unit interval. A family of fuzzy sets defined in $\mathbf{X}$ is denoted by $F(\mathbf{X})$.

Probability-based information granules are expressed in the form of some probability density functions or probability functions. They capture a collection of elements resulting from some experiment. In virtue of the concept of probability, the granularity of information becomes a manifestation of occurrence of some elements.

Rough sets [16–18] emphasize a roughness of description of a given concept X when being realized in terms of the indiscernibility relation provided in advance. The roughness of the description of X is manifested in terms of its lower and upper approximations of a certain rough set. A family of fuzzy sets defined in **X** is denoted by $R(\mathbf{X})$.

Shadowed sets [21] offer a description of information granules by distinguishing among elements, which fully belong to the concept, are excluded from it and whose belongingness is completely unknown. Formally, these information granules are described as a mapping X: $\mathbf{X} \rightarrow \{1, 0, [0,1]\}$ where the elements with the membership quantified as the entire [0,1] interval are used to describe a shadow of the construct. Given the nature of the mapping here, shadowed sets can be sought as a granular description of fuzzy sets where the shadow is used to localize partial membership values, which in fuzzy sets are distributed over the entire universe of discourse. A family of fuzzy sets defined in **X** is denoted by $S(\mathbf{X})$.

Probability-grounded sets are defined over a certain universe where the membership grades are represented as some probabilistic constructs. For instance, each element of a set comes with a truncated to [0,1] probability density function, which quantifies a degree of membership to the information granule. There are a number of variations of these constructs with probabilistic sets [11] being one of them.

Other formal models of information granules involve axiomatic sets, soft sets, and intuitionistic sets.

6.1 Information Granules of Higher Type and Higher Order

In general, we distinguish between information granules of higher type and higher order.

Higher type information granules. The quantification of levels of belongingness to a given information granule is granular itself rather than numeric as encountered in sets or fuzzy sets. This type of quantification is of interest in situations it is not quite justifiable or technically sound to quantify the membership in terms of a single numeric value. These situations give rise to ideas of type-2 fuzzy sets or interval-valued fuzzy sets. In the first case the membership is quantified by a certain fuzzy set taking on values in the unit interval. In the second case we have a subinterval of [0,1] representing membership values. One can discuss fuzzy sets of higher type in which the granular quantification is moved to the higher levels of the construct. For instance, one can talk about type-3, type-4, … fuzzy sets. Albeit conceptually sound, one should be aware that the computing overhead associated with further processing of such information granules becomes more significant. In light of the essence of these constructs, we can view probabilistic granules to be treated as higher type information granules as we admit membership values to be granulated in a probabilistic manner.

Higher order information granules. The notion of higher order of information granules points at a space in which an information granule is defined. Here the universe of discourse is composed of a family of information granules. For instance, a fuzzy set of order 2 is constructed in the space of a family of so-called reference fuzzy sets. This stands in a sharp contrast with fuzzy sets of order 1, which are

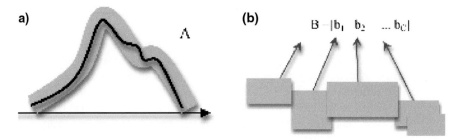

Fig. 1 Examples of information granules of type-2 (type-2 fuzzy set A) contrasted with a fuzzy set (**a**) and order 2 fuzzy set B (defined over a collection of information granule forming the universe of discourse) (**b**)

defined in individual elements of the universe of discourse. One could remark that fuzzy modeling quite often involve order 2 fuzzy sets.

The illustration of these concepts is included in Fig. 1.

These types of construct could be generalized by invoking a number of consecutive levels of the structure. In all situations, we could assess whether moving to the higher level or order constructs is legitimate from the perspective of the problem at hand.

6.2 Hybrid Models of Information Granules

Information granules can embrace several granulation formalisms at the same time forming some hybrid models. This constructs become of particular interest when information granules have to capture a multifaceted nature of the problem. There are a large number of interesting options here. Some of them, which have been found convincing concern.

(a) fuzzy probabilities. Probability and fuzzy sets are orthogonal concepts and as such they could be considered together as a single entity. The concepts of a fuzzy event and fuzzy probabilities (viz. probabilities whose values are quantified in terms of fuzzy sets, say high probability, very low probability) are of interest here.

(b) fuzzy rough and rough fuzzy information granules. Here the indiscernibility relation can be formed on a basis of fuzzy sets. Fuzzy sets, rather than sets are also the entities that are described in terms of the elements of the indiscernibility relation. The original object X for which a rough set is formed might be a fuzzy set itself rather than a set used in the original definition of rough sets.

7   The Concept of Information Granulation-Degranulation

When it comes to numeric information **x** forming a vector in a certain multidimensional space, we can develop an interesting granulation-degranulation scheme [25]. We assume that the information granules forming a collection (codebook) A are described by their prototypes $\mathbf{v}_1$, $\mathbf{v}_2$, ..., $\mathbf{v}_c$. Such prototypes can be formed as a result of fuzzy clustering [4, 12, 24]. The granulation-degranulation task is formulated as a certain optimization problem. In what follows, we assume that the distance used in the solutions is the Euclidean one. The granulation of **x** returns its representation in terms of the collection of available information granules expressed in terms of their prototypes. More specifically, **x** is expressed in the form of the membership grades u_i of **x** to the individual granules A_i, which form a solution to the following optimization problem

$$\text{Min} \sum_{i=1}^{c} u_i^m(\mathbf{x})||\mathbf{x} - \mathbf{v}_i||^2 \tag{1}$$

subject to the following constraints imposed on the degrees of membership

$$\sum_{i=1}^{c} u_i(\mathbf{x}) = 1 \quad u_i(\mathbf{x}) \in [0, 1] \tag{2}$$

where "m" stands for the so-called fuzzification coefficient, m > 1 [4]. The derived solution to the problem above reads as follows

$$u_i(\mathbf{x}) = \frac{1}{\sum_{j=1}^{c} \left(\frac{||\mathbf{x}-\mathbf{v}_i||}{||\mathbf{x}-\mathbf{v}_j||}\right)^{2/(m-1)}} \tag{3}$$

For the degranulation phase, given $u_i(\mathbf{x})$ and the prototypes $\mathbf{v}_i$, the vector $\hat{\mathbf{x}}$ is considered as a solution to the minimization problem in which we reconstruct (degranulate) original **x** when using the prototypes and the membership grades

$$\sum_{i=1}^{c} u_i^m(\mathbf{x})||\hat{\mathbf{x}} - \mathbf{v}_i||^2 \tag{4}$$

Because of the use of the Euclidean distance in the above performance index, the calculations here are straightforward yielding the result

$$\hat{\mathbf{x}} = \frac{\sum_{i=1}^{c} u_i^m(\mathbf{x})\mathbf{v}_i}{\sum_{i=1}^{c} u_i^m(\mathbf{x})} \tag{5}$$

It is important to note that the description of **x** in more abstract fashion realized by means of A_i and being followed by the consecutive degranulation brings about a certain granulation error (which is inevitable given a fact that we move back and forth between different levels of abstraction). While the above formulas pertain to the granulation realized by fuzzy sets, the granulation-degranulation error is also present when dealing with sets (intervals). In this case we are faced with a quantization error, which becomes inevitable when working with A/D (granulation) and D/A (degranulation) conversion mechanisms.

8 Clustering as a Means of Design of Information Granules

Clustering is a commonly encountered way of forming information granules. In objective function-based clustering there is usually a certain constraint imposed on the relationship between the resulting information granules. For instance, one requires that the union of information granules "covers" the entire data set, that is $\bigcup_{i=1}^{c} A_c = D$. Obviously the union operation has to be specified in accordance to the formalism of information granules used there. There are a large number of clustering methods and depending on the formalism being used we end up with the granules expressed in the language of sets, $P(X)$, fuzzy sets $F(X)$, rough sets $R(X)$, shadowed sets $S(X)$ and others. The form of the granules depends on the clustering algorithm and the formulation of the objective function (and partition matrix, in particular). The number of information granules has been another point of comprehensive study as this pertains to the problem of cluster validity.

By having a quick look at the plethora of clustering methods one can conclude that they predominantly realize the concept of closeness between elements: data, which are close to each other form the same information granule. There is another aspect of functional resemblance and this facet is captured through so-called knowledge-based clustering, cf. [24].

8.1 Unsupervised Learning with Fuzzy Sets

Unsupervised learning, quite commonly treated as an equivalent of clustering is aimed at the discovery of structure in data and its representation in the form of clusters—groups of data.

In reality, clusters, in virtue of their nature, are inherently *fuzzy*. Fuzzy sets constitute a natural vehicle to quantify strength of membership of patterns to a certain group. An example shown in Fig. 2 clearly demonstrates this need. The pattern positioned in-between the two well structured and compact groups exhibit some level of resemblance (membership) to each of the clusters. Surely enough, one could be hesitant to allocate it fully to either of the clusters. The membership values

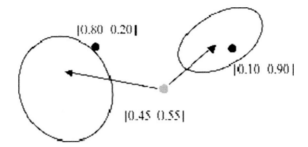

|0.80 0.20|

|0.10 0.90|

|0.45 0.55|

Fig. 2 Example of two-dimensional data with patterns of varying membership degrees to the two highly visible and compact clusters

such as e.g., 0.55 and 0.45 are not only reflective of the structure in the data but they flag (highlight) the distinct nature of this data—and maybe trigger some further inspection of this pattern. In this way we remark a user-centric character of fuzzy sets, which make interaction with users more effective and transparent.

8.2   Fuzzy C-Means as an Algorithmic Vehicle of Data Reduction Through Fuzzy Clusters

Fuzzy sets can be formed on a basis of numeric data through their clustering (groupings). The groups of data give rise to membership functions that convey a global more abstract and general view at the available data. With this regard Fuzzy C-Means (FCM, for brief) is one of the commonly used mechanisms of fuzzy clustering [24].

Let us review its formulation, develop the algorithm and highlight the main properties of the fuzzy clusters. Given a collection of n-dimensional data set $\{x_k\}$, $k = 1,2,...,N$, the task of determining its structure—a collection of "c" clusters, is expressed as a minimization of the following objective function (performance index) Q being regarded as a sum of the squared distances between data and their representatives (prototypes)

$$Q = \sum_{i=1}^{c} \sum_{k=1}^{N} u_{ik}^{m} ||x_k - v_i||^2 \tag{6}$$

Here v_i s are n-dimensional prototypes of the clusters, $i = 1, 2,..., c$ and $U = [u_{ik}]$ stands for a partition matrix expressing a way of allocation of the data to the corresponding clusters; u_{ik} is the membership degree of data x_k in the i-th cluster. The distance between the data z_k and prototype v_i is denoted by $||.||$. The fuzzification coefficient m (>1.0) expresses the impact of the membership grades on the individual clusters. It implies as certain geometry of fuzzy sets. A partition matrix satisfies two important and intuitively appealing properties

$$\text{(a)} \quad 0 < \sum_{k=1}^{N} u_{ik} < N, \quad i = 1, 2, \ldots, c$$

$$\text{(b)} \quad \sum_{i=1}^{c} u_{ik} = 1, \quad k = 1, 2, \ldots, N \tag{7}$$

Let us denote by U a family of matrices satisfying (a)–(b). The first requirement states that each cluster has to be nonempty and different from the entire set. The second requirement states that the sum of the membership grades should be confined to 1.

The minimization of Q completed with respect to $U \in \mathbf{U}$ and the prototypes $\mathbf{v}_i$ of $V = \{\mathbf{v}_1, \mathbf{v}_2, \ldots \mathbf{v}_c\}$ of the clusters. More explicitly, we write it down as follows

$$\min Q \quad \text{with respect to } U \in \mathbf{U}, \mathbf{v}_1, \mathbf{v}_2, \ldots, \mathbf{v}_c \in \mathbf{R}^n \tag{8}$$

The successive entries of the partition matrix are expressed as follows

$$u_{st} = \frac{1}{\sum_{j=1}^{c} \left(\frac{||\mathbf{x}_t - \mathbf{v}_s||}{||\mathbf{x}_t - \mathbf{v}_j||} \right)^{2/(m-1)}} \tag{9}$$

Assuming the Euclidean form of distance, the prototypes $\mathbf{v}_1, \mathbf{v}_2, \ldots, \mathbf{v}_c$ come in the form

$$\mathbf{v}_s = \frac{\sum_{k=1}^{N} u_{ik}^m \mathbf{x}_k}{\sum_{k=1}^{N} u_{ik}^m} \tag{10}$$

In addition to the Euclidean distance studied are different distances, say the Hamming or the Tchebyschev one [20]. Overall, the FCM clustering is completed through a sequence of iterations where we start from some random allocation of data (a certain randomly initialized partition matrix) and carry out the following updates by adjusting the values of the partition matrix and the prototypes. The iterative process is continued until a certain termination criterion has been satisfied. Typically, the termination condition is quantified by looking at the changes in the membership values of the successive partition matrices. Denote by $U(t)$ and $U(t + 1)$ the two partition matrices produced in the two consecutive iterations of the algorithm. If the distance $||U(t + 1) - U(t)||$ is les than a small predefined threshold ε (say, $\varepsilon = 10^{-5}$ or 10^{-6}), then we terminate the algorithm. Typically, one considers the Tchebyschev distance between the partition matrices meaning that the termination criterion reads as follows

$$\max_{i,k} |u_{ik}(t + 1) - u_{ik}(t)| \leq \varepsilon \tag{11}$$

8.3 Knowledge-Based Clustering

Clustering and classification are positioned at the two opposite poles of the learning paradigm. In reality, there is no "pure" unsupervised learning as usually there is some limited amount of domain knowledge. There is no fully supervised learning as some labels might not be completely reliable (as those encountered in case of learning with probabilistic teacher).

There is some domain knowledge and it has to be carefully incorporated into the generic clustering procedure. Knowledge hints can be conveniently captured and formalized in terms of fuzzy sets. Altogether with the underlying clustering algorithms, they give rise to the concept of knowledge-based clustering—a unified framework in which data and knowledge are processed together in a uniform fashion.

We can distinguish several interesting and practically viable ways in which domain knowledge is taken into consideration:

A subset of labeled patterns The knowledge hints are provided in the form of a small subset of labeled patterns $K \subset N$ [22, 23]. For each of them we have a vector of membership grades f_k, $k \in K$ which consists of degrees of membership the pattern is assigned to the corresponding clusters. As usual, we have $f_{ik} \in [0, 1]$ and $\sum_{i=1}^{c} f_{ik} = 1$.

Proximity-based clustering Here we are provided a collection of pairs of patterns [13] with specified levels of closeness (resemblance) which are quantified in terms of proximity, prox(k, l) expressed for x_k and x_l. The proximity offers a very general quantification scheme of resemblance: we require reflexivity and symmetry, that is prox(k, k) = 1 and prox(k, l) = prox(l, k) however no transitivity is needed.

"belong" and "not-belong" Boolean relationships between patterns These two Boolean relationships stress that two patterns should belong to the same clusters, $R(x_k, x_l) = 1$ or they should be placed apart in two different clusters, $R(x_k, x_l) = 0$. These two requirements could be relaxed by requiring that these two relationships return values close to one or zero.

Uncertainty of labeling/allocation of patterns We may consider that some patterns are "easy" to assign to clusters while some others are inherently difficult to deal with meaning that their cluster allocation is associated with a significant level of uncertainty. Let $F(x_k)$ stands for the uncertainty measure (e.g., entropy) for x_k (as a matter of fact, F is computed for the membership degrees of x_k that is $F(u_k)$ with u_k being the kth column of the partition matrix. The uncertainty hint is quantified by values close to 0 or 1 depending upon what uncertainty level a given pattern is coming from.

Depending on the character of the knowledge hints, the original clustering algorithm needs to be properly refined. In particular the underlying objective function has to be augmented to capture the knowledge-based requirements. Below shown are several examples of the extended objective functions dealing with the knowledge hints introduced above.

When dealing with some labeled patterns we consider the following augmented objective function

$$Q = \sum_{i=1}^{c} \sum_{k=1}^{N} u_{ik}^{m} ||\mathbf{x}_k - \mathbf{v}_i||^2 + \alpha \sum_{i=1}^{c} \sum_{k=1}^{N} (u_{ik} - f_{ik} b_k)^2 ||\mathbf{x}_k - \mathbf{v}_i||^2 \qquad (12)$$

where the second term quantifies distances between the class membership of the labeled patterns and the values of the partition matrix. The positive weight factor (a) helps set up a suitable balance between the knowledge about classes already available and the structure revealed by the clustering algorithm. The Boolean variable b_k assumes values equal to 1 when the corresponding pattern has been labeled.

The proximity constraints are accommodated as a part of the optimization problem where we minimize the distances between proximity values being provided and those generated by the partition matrix $P(k_1, k_2)$

$$Q = \sum_{i=1}^{c} \sum_{k=1}^{N} u_{ik}^{m} ||\mathbf{x}_k - \mathbf{v}_i||^2$$

$$||\text{prox}(k_1, k_2) - P(k_1, k_2)|| \rightarrow \text{Min } k_1, \ k_2 \in K \qquad (13)$$

with K being a pair of patterns for which the proximity level has been provided. It can be shown that given the partition matrix the expression $\sum_{i=1}^{c} \min(u_{ik1}, u_{ik2})$ generates the corresponding proximity value.

For the uncertainty constraints, the minimization problem can be expressed as follows

$$Q = \sum_{i=1}^{c} \sum_{k=1}^{N} u_{ik}^{m} ||\mathbf{x}_k - \mathbf{v}_i||^2$$

$$||F(\mathbf{u}_k) - g_k|| \rightarrow \text{Min } k \in K \qquad (14)$$

where K stands for the set of patterns for which we are provided with the uncertainty values g_k.

Undoubtedly the extended objective functions call for the optimization scheme that is more demanding as far as the calculations are concerned. In several cases we cannot modify the standard technique of Lagrange multipliers, which leads to an iterative scheme of successive updates of the partition matrix and the prototypes. In general, though, the knowledge hints give rise to a more complex objective function in which the iterative scheme cannot be useful in the determination of the partition matrix and the prototypes. Alluding to the generic FCM scheme, we observe that the calculations of the prototypes in the iterative loop are manageable in case of the Euclidean distance. Even the Hamming or Tchebyshev distance brings a great deal of complexity. Likewise, the knowledge hints lead to the increased complexity: the prototypes cannot be computed in a straightforward way and one has to resort himself to more advanced optimization techniques. Evolutionary computing arises here as an appealing alternative. We may consider any of the options available there including genetic algorithms, particle swarm optimization, ant colonies, to name some of them. The general scheme can be schematically structured as follows:

- repeat {EC (prototypes); compute partition matrix U;}

  One can recall also studies in the formation of granular prototypes [2].

9 Computational Intelligence and Software Engineering

Software Engineering with its scope of challenges and open problems whose formulation can be expressed in the setting of CI. CI can offer some viable solutions to the problem. There are several essential facets to be raised here:

Quality of data In software processes and software artifacts, we envision existence of data at different levels of abstraction (granularity). There might be data expressed numerically (coming as a results of measuring). There are pieces of data expressed by experts and those could be expressed as outcomes of questionnaires and because of this, they might be expressed in linguistic terms. Some initial documents of software requirements could include linguistic rather than numeric quantification. The data could be incomplete and call for mechanisms of data imputation.

In view of the nature of the categories of the problems listed above, one can stress that Granular Computing (fuzzy sets, rough sets, interval analysis) deliver a sound solution both in terms of the underlying methodology and the ensuing algorithms. The processing of information granules regarded as input variables of the models leads to the results that are again expressed as information granules (by adhering to the ideas of propagation of information granularity). Some optimization activities required here could be supported by population-based methods of CI.

Structure of models Software Engineering is not subject to laws of physics and does not adhere to general and quantifiable general laws. For instance, this manifests very profoundly in software cost estimation. While COCOMO-like models offer sound alternative, it becomes quite apparent that the accuracy of any model is limited and intensive calibrating of the model becomes a necessity. There is a great deal of diversity of software systems, process, quality indexes and each software project calls for a quite different model to form its proper and efficient description. Here the three pillars of CI assume an essential role. Neural networks serve as adaptive and nonlinear models whose flexibility arises as an important asset in capturing the essence of available data. Information granules help accommodate non-numeric data, which becomes crucial to evaluate the quality of the results and create a feedback as to the refinement (specialization) of available experimental evidence.

The complexity of the models and a genuine requirement to optimize them both structurally and parametrically emphasizes a need for the involvement of population-based optimization to construct the models.

Interpretation of results In light of propagation of information granularity, the results of processing are information granules. Their use creates an evident advantage by delivering an efficient means to quantify the precision of the results depending upon the granularity of the available input data. The granularity of the results makes them to become in a stronger rapport with the reality as to the specificity of obtained findings.

10   Conclusions

We have outlined the fundamentals of Computational Intelligence showing that the synergy of the technologies of fuzzy sets becomes a vital component in the analysis of software data and design models of software artifacts.

With this regard, fuzzy sets or being more general, information granules, form an important front- and back-end of these constructs. By forming the front end, they help develop a suitable view at the software data, incorporate available domain knowledge and come up with a feature space that supports the effectiveness of ensuing processing, quite commonly engaging various schemes of neurocomputing or evolutionary neurocomputing. Equally important role is played by fuzzy sets in the realization of the back end of the overall processing scheme: they strengthen the interpretability of results of data analysis or outcomes produced by models of software processes and systems.

References

1. Bargiela, A., Pedrycz, W.: Recursive information granulation: aggregation and interpretation issues. IEEE Trans. Syst. Man Cybern. B **33**(1), 96–112 (2003)
2. Bargiela, A., Pedrycz, W., Hirota, K.: Granular prototyping in fuzzy clustering. IEEE Trans. Fuzzy Syst. **12**(5), 697–709 (2004)
3. Bargiela, A., Pedrycz, W.: Granular Computing: An Introduction. Kluwer Academic Publishers, Dordrecht (2002)
4. Bezdek, J.C.: Pattern Recognition with Fuzzy Objective Function Algorithms. Plenum Press, New York (1981)
5. Bezdek, J.C.: On the relationship between neural networks, pattern recognition and intelligence. Int. J. Approx. Reason. **6**(2), 85–107 (1992)
6. Bezdek, J.C.: What is computational intelligence. In: Robinson, C.J., Zurada, J.M., Marks II, R.J. (eds.) Computational Intelligence Imitating Life, pp. 1–12. IEEE Press, Piscataway, NJ (1994)
7. Engelbrecht, A.P.: Fundamentals of Computational Swarm Intelligence. Wiley, London, UK (2005)

8. Fulcher, J., Jain, L.C. (eds): Computational Intelligence: A Compendium. Springer, Berlin (2008)
9. Goldberg, D.E.: Genetic Algorithms in Search, Optimization, and Machine Learning. Addison Wesley, Reading, MA (1989)
10. Haykin, S.: Neural Networks: A Comprehensive Foundation, 2nd edn. Prentice Hall Upper Saddle River, NJ (1999)
11. Hirota, K.: Concepts of probabilistic sets. Fuzzy Sets Syst. 5(1), 31–46 (1981)
12. Hoppner, F., et al.: Fuzzy Cluster Analysis. Wiley, Chichester (1999)
13. Loia, V., Pedrycz, W., Senatore, S.: P-FCM: a proximity-based fuzzy clustering for user-centered web applications. Int. J. Approx. Reason. 34, 121–144 (2003)
14. Moore, R.: Interval Analysis. Prentice-Hall, Englewood Cliffs, NJ (1966)
15. Mumford, C.L., Jain, L.C. (eds.): Computational Intelligence. Springer, Berlin (2009)
16. Pawlak, Z.: Rough sets. Int. J. Comput. Inf. Sci. 11, 341–356 (1982)
17. Pawlak, Z.: Rough Sets. Theoretical Aspects of Reasoning About Data. Kluwer Academic Publishers, Dordercht (1991)
18. Pawlak, Z., Skowron, A.: Rough sets and Boolean reasoning. Inf. Sci. 177(1), 41–73 (2007)
19. Pedrycz, W., Bargiela, A.: Granular clustering: a granular signature of data. IEEE Trans. Syst. Man Cybern. 32(2), 212–224 (2002)
20. Pedrycz, W., Bargiela, A.: A model of granular data: a design problem with the Tchebyschev FCM. Soft. Comput. 9(3), 155–163 (2005)
21. Pedrycz, W.: Shadowed sets: representing and processing fuzzy sets. IEEE Trans. Syst. Man Cybern Part B 28, 103–109 (1998)
22. Pedrycz, W., Waletzky, J.: Neural network front-ends in unsupervised learning. IEEE Trans. Neural Netw 8, 390–401 (1997)
23. Pedrycz, W., Waletzky, J.: Fuzzy clustering with partial supervision. IEEE Trans. Syst. Man Cybern. 5, 787–795 (1997)
24. Pedrycz, W.: Knowledge-Based Clustering: From Data to Information Granules. J. Wiley, Hoboken, NJ (2005)
25. Pedrycz, W., Gomide, F.: Fuzzy Systems Engineering. Wiley, Hoboken, NJ (2007)
26. Pedrycz, W.: Computational Intelligence: An Introduction. CRC Press, Boca Raton, Fl (1997)
27. Wassermann, P.D.: Neural Computing: Theory and Practice. Van Nostrand, Reinhold, New York, NY (1989)
28. Zadeh, L.A.: Fuzzy sets. Inf. Control 8, 338–353 (1965)
29. Zadeh, L.A.: Towards a theory of fuzzy information granulation and its centrality in human reasoning and fuzzy logic. Fuzzy Sets Syst. 90, 111–117 (1997)
30. Zadeh, L.A.: Toward a generalized theory of uncertainty (GTU)—An outline. Inf. Sci. 172, 1–40 (2005)

Towards Benchmarking Feature Subset Selection Methods for Software Fault Prediction

Wasif Afzal and Richard Torkar

Abstract Despite the general acceptance that software engineering datasets often contain noisy, irrelevant or redundant variables, very few benchmark studies of feature subset selection (FSS) methods on real-life data from software projects have been conducted. This paper provides an empirical comparison of state-of-the-art FSS methods: information gain attribute ranking (IG); Relief (RLF); principal component analysis (PCA); correlation-based feature selection (CFS); consistency-based subset evaluation (CNS); wrapper subset evaluation (WRP); and an evolutionary computation method, genetic programming (GP), on five fault prediction datasets from the PROMISE data repository. For all the datasets, the area under the receiver operating characteristic curve—the AUC value averaged over 10-fold cross-validation runs—was calculated for each FSS method-dataset combination before and after FSS. Two diverse learning algorithms, C4.5 and naïve Bayes (NB) are used to test the attribute sets given by each FSS method. The results show that although there are no statistically significant differences between the AUC values for the different FSS methods for both C4.5 and NB, a smaller set of FSS methods (IG, RLF, GP) consistently select fewer attributes without degrading classification accuracy. We conclude that in general, FSS is beneficial as it helps improve classification accuracy of NB and C4.5. There is no single best

W. Afzal (✉)
School of Innovation, Design & Engineering, Mälardalen University,
Västerås, Sweden
e-mail: wasif.afzal@mdh.se

R. Torkar
Blekinge Institute of Technology, Karlskrona, Sweden
e-mail: richard.torkar@cse.gu.se

R. Torkar
Chalmers University of Technology, Gothenburg, Sweden

R. Torkar
University of Gothenburg, Gothenburg, Sweden

W. Afzal
Department of Computer Science, Bahria University, Islamabad, Pakistan

© Springer International Publishing Switzerland 2016
W. Pedrycz et al. (eds.), *Computational Intelligence and Quantitative Software Engineering*, Studies in Computational Intelligence 617,
DOI 10.1007/978-3-319-25964-2_3

FSS method for all datasets but IG, RLF and GP consistently select fewer attributes without degrading classification accuracy within statistically significant boundaries.

Keywords Feature subset selection · Fault prediction · Empirical

1 Introduction

A bulk of literature on prediction and estimation in software engineering contributes to software fault/defect prediction (also termed as software quality classification/ software quality modeling). Software fault prediction research uses software metrics to predict the response variable which can either be the class of a module (e.g., fault-prone and not fault-prone) or a quality factor (e.g., number of faults) for a module [1]. This paper is concerned with classifying software components/modules as fault-prone and not fault-prone [2, 3, 4]. Such a classification task is useful for the following reasons:

- Knowing which software components are likely to be fault-prone supports better targeting of software testing effort. This in turn has the potential to improve test efficiency and effectiveness.
- Fault-prone software components are candidates of refactoring whereby their internal structure can be improved.

Despite the presence of a large number of models for software fault prediction, there is lack of a definitive advice on what prediction models are useful under different contexts. In order to increase confidence in the results of software fault prediction studies, more and more research is focussing on the need for a robust process and methodology to build prediction models [5, 3, 6]. Central to such a methodology are issues such as data quality, measurement of predictive performance [7], resampling methods to use [8] and reporting of fault prediction experiments [3]. For data quality, important issues are data preprocessing [9], class imbalance [10, 11] and impact of feature subset selection (FSS) methods [12, 13]. This paper contributes to the last aspect of data quality: use of FSS methods in software fault prediction.

The purpose of FSS is to find a subset of the original features of a dataset, such that an induction algorithm that is run on data containing only these features generates a classifier with the highest possible accuracy [14]. There are several reasons to keep the number of features in a data set as small as possible:

1. Reducing the number of features allows classification algorithms to operate faster, more effectively [15] and with greater simplicity [16].
2. Smaller number of features help reduce the curse of dimensionality.[1]

---

[1]The requirement that the number of training data points to be an exponential function of the feature dimension.

3. Smaller number of features reduce measurement cost as less data needs to be collected [17].
4. FSS helps to achieve a better understandable model and simplifies the usage of various visualization techniques [18].

The simplest approach to FSS would require examining all possible subsets of the desired number of features in the selected subset and then selecting the subset with the smallest classification error. However, this leads to a combinatorial explosion, making exhaustive search all but impractical for most of the data sets [16]. Naturally many FSS methods are search-based [19], combined with an attribute utility estimator to evaluate the relative merit of alternate subsets of attributes [15].

Several researchers in software engineering have emphasized the need to investigate only relevant variables. According to Dybå et al. [20]: "*Careful selection of which independent variables to include and which variables to exclude, is, thus, crucial to raising the power of a study and the legitimacy of its potential findings*". Further emphasizing the importance of FSS, Song et al. [6] argue that "*[…] before building prediction models, we should choose the combination of all three of learning algorithm, data pre-processing and attribute selection method, not merely one or two of them*". It is also generally accepted that software engineering data sets often contain noisy, irrelevant, or redundant variables [17, 21], therefore, it is important to evaluate FSS methods for software engineering data sets. In literature, there are few studies that compare FSS methods for software fault prediction (Sect. 2) but no benchmark study on commonly used FSS methods on real-life public data from software projects has been conducted. Moreover, the use of evolutionary algorithms (e.g., genetic algorithm, genetic programming) have sporadically been investigated as FSS methods [22, 23, 24, 25] but not to an extent of comparing with state-of-the-art FSS methods, using publicly available real-life data from software projects.

This paper provides an empirical comparison of the state-of-the-art FSS methods and an evolutionary computation method (genetic programming (GP)) on five software fault prediction data sets from the PROMISE data repository [26]. Two diverse learning algorithms, C4.5 and naïve Bayes (NB), are used to test the attribute sets given by each FSS method. We are interested in investigating if the classification accuracy of C4.5 and NB significantly differ before and after the application of FSS methods. In order to formalize the purpose of the empirical study, we set forth the following hypotheses to test:

H_0: The classification accuracy of C4.5 and NB is not significantly different before and after applying the FSS methods, i.e., $ACC_{C4.5} = ACC_{NB}$.
H_1: The classification accuracy of C4.5 and NB is significantly different before and after applying the FSS methods, i.e., $ACC_{C4.5} \neq ACC_{NB}$.

The results of this study indicate that FSS is generally useful for software fault prediction using NB and C4.5. However there are no clear winners for either of the two learning algorithms for the variety of FSS methods used. Based on individual accuracy values, RLF, IG and GP are the best FSS methods for software fault prediction accuracy while CNS and CFS are also good overall performers.

The paper is organized as follows. The next Section describes related work. Section 3 describes the FSS methods used in this study while Sect. 4 describes the datasets used, the evaluation measure and the experimental setup of the replication study. Section 5 presents the results of the empirical study and presents a discussion. Validity evaluation is given in Sect. 6 while the paper is concluded in Sect. 7.

2 Related Work

Molina et al. [27], Guyon and Elisseeff [28], Blum and Langley [29], Dash and Liu [30] and Liu and Yu [31] provide good surveys reviewing work in machine learning on FSS. This section will, however, summarize the work done on FSS in predictive modeling within software engineering. FSS techniques in software engineering have been applied for software cost/effort estimation and software quality classification (also called as software defect/fault prediction). As the below paragraphs would illustrate, there is no definitive guidance available on FSS techniques to use in software engineering predictive modeling.

Dejaeger et al. [32] used a generic backward input selection wrapper for FSS and reported significantly improved performance for software cost modeling in comparison to when no FSS was used. Similar results were reported by Chen et al. [17, 33]. They showed that using wrapper improves software cost prediction accuracy and is further enhanced when used in combination with row pruning. However, for the COCOMO-styled datasets used in a study by Menzies et al. [34] for estimating software effort/cost, wrapper FSS technique did not improve the estimation accuracy. Kirsopp et al. [35] used random seeding, hill climbing and forward sequential selection to search for optimal feature subsets for predicting software project effort. They showed that hill climbing and forward sequential selection produce better results than random searching. Azzeh et al. [36], on the other hand, showed that their proposed fuzzy FSS algorithm consistently outperforms hill climbing, forward subset selection and backward subset selection for software effort estimation. Li et al. [37] showed that a hybrid of wrapper and filter FSS techniques known as mutual information based feature selection (MICBR) can select more meaningful features while the performance was comparable to exhaustive search, hill climbing and forward sequential selection.

Menzies et al. [38] showed that there are no clear winners in FSS techniques for learning defect predictors for software fault/defect prediction. They compared information gain, correlation-based feature selection, relief and consistency based subset evaluation. Rodriguez et al. [12, 13] also compared filter and wrapper FSS techniques for predicting faulty modules. They, however, concluded that wrapper FSS techniques have better accuracy than filter FSS techniques. Song et al. [6] used wrapper FSS with forward selection and backward elimination search strategies for defect proneness prediction. They showed that different attribute selectors are suitable to different learning algorithms. Catal and Diri [39] applied correlation-based FSS method on class-level and method-level metrics for software fault prediction. They showed that random forests gives the best results when using this FSS method. Khoshgoftaar et al. [40] found that the use of a stepwise regression model and a correlation-based FSS with greedy forward search did not yield improved predictions. Wang et al. [41] compared seven filter based FSS techniques and proposed their own combination of filter-based and consistency-based FSS algorithm. Their proposed algorithm and the Kolmogorov-Smirnov technique performed competitively with other FSS techniques. Koshgoftaar et al. [42] also showed better results with a FSS method based on the Kolmogorov-Smirnov two-sample statistical test. Altidor et al. [43] compared their new wrapper FSS algorithm against 3-fold cross-validation, 3-fold cross-validation risk impact and a combination of the two. They showed that the performance of their new FSS technique is dependent on the base classifier (ranker aid), the performance metric and the methodology. Gao et al. [44] concluded that data sampling followed by wrapper FSS technique improves the accuracy of predicting high-risk program modules. Gao et al. [45] compared seven feature ranking techniques and four FSS techniques. Their proposed automatic hybrid search performed best among FSS techniques. Khoshgoftaar et al. [46] compared seven filter-based feature ranking techniques, including a signal-to-noise (SNR) technique. SNR performed as well as the best performer of the six commonly used techniques. Wang et al. [47] compared several ensemble FSS techniques and concluded that although there are no clear winners but ensembles of few rankers are effective then ensembles of many rankers. Khoshgoftaar et al. [10] investigated the relation between six filter-based FSS methods with random under-sampling technique. They concluded that FSS based on sampled data resulted in significantly better performance than FSS based on original data.

3   Feature Subset Selection (FSS) Methods

There are two commonly known categories of FSS methods: the filter approach and the wrapper approach. In the filter approach, the feature selection takes place independently of the learning algorithm and is based only on the data characteristics. The wrapper approach, on the other hand, conducts a search for a good subset using the learning algorithm itself as part of the evaluation function [14, 15] provides another categorization for FSS methods, namely, those methods that evaluate individual attributes and those that evaluate subset of attributes.

We have chosen to empirically evaluate a total of seven FSS methods, two that evaluate individual attributes (information gain attribute ranking and Relief), three that evaluate subsets of attributes (correlation-based feature selection, consistency-based subset evaluation, wrapper subset evaluation), one classical statistical method for dimensionality reduction (principal components analysis) and

one evolutionary computational method (genetic programming). Following is a
brief description of the FSS methods used in this study.

3.1 Information Gain (IG) Attribute Ranking

The foundation of IG attribute ranking is the concept of entropy which is considered
as a measure of system's unpredictability. If C is the class, the entropy of C is
given by:

$$H(C) = -\sum p(c) \log p(c)$$

where $p(c)$ is the marginal probability density function for class C. If the observed
values of C are partitioned based on an attribute A and the entropy of C after
observing the attribute is less than the entropy of C prior to it, there is a relationship
between C and A. The entropy of C after observing A is:

$$H(C|A) = -\sum_{a \in A} p(a) \sum_{c \in C} p(c|a) \log p(c|a)$$

where $p(c|a)$ is the conditional probability of c given a.

Given that entropy is a measure of system's unpredictability, information gain is
the amount by which the entropy of C decreases [48]. It is given by:

$$IG = H(C) - H(C|A) = H(A) - H(A|C)$$

IG is a symmetrical measure meaning that information gained about C after
observing A is equal to the information gained about A after observing C.

IG attribute ranking is one of the simplest and fastest attribute ranking methods
[15] but its weakness is that it is biased in favor of attributes with more instances
even when they are not more informative [49]

In this study, IG attribute ranking is used with the ranker search method that
ranks attributes by their individual evaluations.

3.2 Relief (RLF)

Relief is an instance-based attribute raking algorithm proposed by Kira and Rendell
[50]. It estimates the quality of attributes according to how they differentiate
between instances from different classes that are near to each other. So given a
randomly selected instance R, Relief searches for a nearest hit H (a nearest neighbor
from the same class) and a nearest miss M (a nearest neighbor from a different
class). It then updates the relevance score for attributes depending on their values

Fig. 1 The basic Relief
algorithm

Algorithm Relief
Input: for each training instance a vector of attribute values and the class value
Output: the vector W of estimations of the qualities of attributes

1.   set all weights W[A] := 0.0;
2.   **for** i := 1 **to** m **do begin**
3.   randomly select an instance R;
4.   find nearest hit H and nearest miss M;
5.   **for** A := 1 **to** #all_attributes **do**
6.   W[A] := W[A] - diff(A,R,H)/m + diff(A,R,M)/m;
7.   **end**;

for *R, M* and *H*. The process is repeated for a user-defined number of instances
m. The basic Relief algorithm, taken from [51], is given in Fig. 1.

The function *diff* (*Attibute, Instance*1, *Instance*2) calculates the difference
between the values of attribute for two instances. For discrete attributes, the difference is either 1 (the values are different) or 0 (the values are the same). For
continuous attributes the difference is the actual difference normalized to the
interval [0, 1] [15].

In this study the Relief method is used with the ranker search method that ranks
attributes by their individual evaluations and the value of *m* is set to 250 which is a
recommended figure [15].

3.3 *Principal Component Analysis (PCA)*

Principal Component Analysis (PCA) is a statistical technique that transforms a set
of possibly correlated variables into a set of linearly uncorrelated variables. These
linearly uncorrelated variables are called principal components. The transformation
is done by first computing the covariance matrix of the original variables and then
finding its Eigen vectors (principal components). The principal components have
the property that most of their information content is stored in the first few features
so that remainder can be discarded. In this study, PCA is used with the ranker
search method that ranks attributes by their individual evaluations.

3.4 *Correlation-Based Feature Selection (CFS)*

Correlation-based feature selection (CFS) evaluates subsets of attributes rather than
individual attributes [52]. The technique uses a heuristic to evaluate subset of
attributes. The heuristic balances how predictive a group of features are and how
much redundancy is among them.

$$Merit_s = \frac{kr_{cf}}{\sqrt{k + k(k-1)r_{ff}}}$$

where $Merit_s$ is the heuristic merit of a feature subset s containing k features, r_{cf} is the average feature-class correlation and r_{ff} is the average feature-feature intercorrelation [15]. In order to apply $Merit_s$, a correlation matrix has to be calculated and a heuristic search to find a good subset of features. In this study, CFS is used with the Greedy stepwise forward search through the space of attribute subsets.

3.5 Consistency-Based Subset Evaluation (CNS)

Consistency-based subset evaluation (CNS) is also an attribute subset selection technique that uses class consistency as an evaluation metric [53]. CNS looks for combinations of attributes whose values divide the data into subsets containing a strong single class majority [15]. Liu and Setiono [53] proposed the following consistency metric:

$$Consistency_s = 1 - \frac{\sum_{i=0}^{j} |D_i| - |Mi|}{N}$$

where s is an attribute subset, j is the number of distinct combinations of attribute values for s, $|D_i|$ is the number of occurrences of the ith attribute value combination, $|M_i|$ is the cardinality of the majority class for the ith attribute value combination and N is the total number of instances in the data set.

In this study, greedy stepwise forward search is used to produce a set of attributes, ranked according to their overall contribution to the consistency of the attribute set [15].

3.6 Wrapper Subset Evaluation (WRP)

The wrapper feature subset evaluation conducts a search for a good subset using the learning algorithm itself as part of the evaluation function. In this study, repeated five-fold cross-validation is used as an estimate for the accuracy of the classifier while a greedy stepwise forward search is used to produce a list of attributes, ranked according to their overall contribution to the accuracy of the attribute set with respect to the target learning algorithm [15].

3.7 Genetic Programming (GP)

Genetic programming (GP) is an evolutionary computation technique and is an extension of genetic algorithms. It is a *"systematic, domain-independent method for getting computers to solve problems automatically starting from a high-level statement of what needs to be done"* [54]. GP applies iterative, random variation to an existing pool of computer programs to form a new generation of programs by applying analogs of naturally occurring genetic operators [19]. The basic steps in a GP system are given below [54]:

1. Randomly create an initial population of programs.
2. Repeat (until stopping criterion is reached):
   - (a) Execute each program and evaluate its fitness.
   - (b) Select one or two programs to undergo genetic operations.
   - (c) Create new programs by applying the genetic operations.
3. Return the best individual.

As compared with genetic algorithms, the population structures (individuals) in GP are not fixed length character strings, but programs that, when executed, are the candidate solutions to the problem.

The evolution of software fault prediction models using GP is an example of a symbolic regression problem. Symbolic regression is an error-driven evolution as it aims to find a function, in symbolic form, that fits (or approximately fits) data from an unknown curve [55]. In simpler terms, symbolic regression finds a function whose output matches some target values. GP is well suited for symbolic regression problems, as it does not make any assumptions about the structure of the function.

Programs are expressed in GP as syntax trees, with the nodes indicating the instructions to execute and are called functions (e.g., *min*, $*$, $+$, $/$), while the tree leaves are called terminals which may consist of independent variables of the problem and random constants (e.g., x, y, 3). The fitness evaluation of a particular individual is determined by the correctness of the logical output produced for all of the fitness cases [55]. The fitness function guides the search in promising areas of the search space and is a way of communicating a problem's requirements to the GP algorithm. The control parameters limit and control how the search is performed like setting the population size and probabilities of performing the genetic operations. The termination criterion specifies the ending condition for the GP run and typically includes a maximum number of generations [19]. GP iteratively transforms a population of computer programs into a new generation of programs using various genetic operators. Typical operators include crossover, mutation and reproduction. It is expected that over successive iterations, more and more useful structures or programs be evolved, eventually resulting in a structure having most useful sub-components. That structure would then represent the optimal or near-optimal solution to the problem. The crossover operator creates new structure (s) by combining randomly chosen parts from two selected programs or structures.

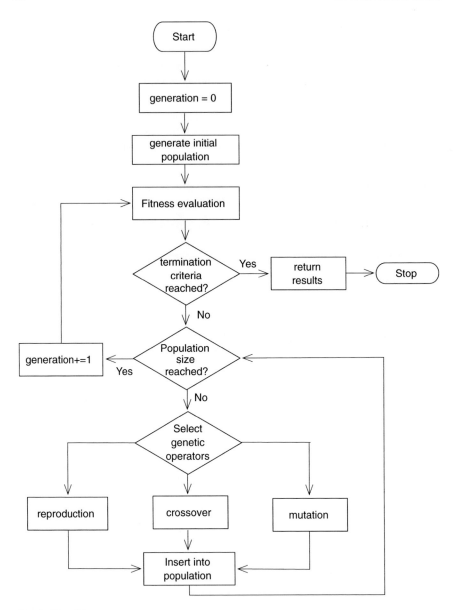

Fig. 2 The GP process

The mutation operator creates a new structure by randomly altering a chosen part of a program. The reproduction operator simply copies a selected structure to the new population. Figure 2 shows the flowchart of the GP process.

For this study, the best GP program (having the minimum $\sum_{i=1}^{n} |e_i - e_i'|$, where e_i is the actual outcome, e_i' is the classification result and n is the size of the data set

Table 1 GP control parameters

Control parameter	Value
Population size	50
Termination condition	500 generations
Function set	$\{+, -, *, /, \sin, \cos, \log, \text{sqrt}\}$
Tree initialization	Ramped half-and-half method
Probabilities of crossover, mutation, reproduction	0.8, 0.1, 0.1
Selection method	roulette-wheel

used to train the GP models) over the 10 runs of each fold of the 10-fold cross-validation is selected. The features making up this best GP program is then designated as the features selected by the GP algorithm. The control parameters that were chosen for the GP system are shown in Table 1. We did not fine tune these parameters for each new data set so as not to bias the results. The population size is related to the size of search space because if the search space is too large, GP will take longer times to find better solutions. The population size was fixed to 50 and this decision was based on our prior experience in experimentation with GP. The termination condition was set to 500 generations and was selected to give enough chance to GP for promoting variety in each generation. The tree initialization method selected was ramped half-and-half which results in very diverse population of trees, with balanced and unbalanced trees of several different depths [56]. The probabilities of crossover, mutation and reproduction were set to 0.8, 0.1 and 0.1 respectively which was done again to promote maximum variation. The selection method used was roulette-wheel which is one of the few sampling methods used in GP to select parent individuals to produce their children.

4 Experimental Setup

In order to compare the performance of different FSS methods, the attribute sets selected by each method are tested with two learning algorithms, namely C4.5 and NB. These algorithms represent two different approaches (C4.5 being a decision-tree learner and NB being a probabilistic learner) and are considered state-of-the-art techniques. Also one of the previous benchmark studies [15] have used the same algorithms for comparing the effectiveness of attribute selection.

The NB classifier is based on the Bayesian theorem. It analyses each data attribute independently and being equally important. The NB classifier learns the conditional probability of each attribute A_i given the class label C, from the training data. Classification is done by applying the Bayes rule to compute the probability of C given the particular instance of $A_1 \ldots An$, and then predicting the class with the highest posterior probability [57]. The NB classifier assumes that features are independent given class, that is, $P(X|C) = \prod_{i=1}^{n} P(X_i|C)$ where $X = (X_1 \ldots X_n)$ is a

Table 2 Characteristics of datasets used in the study

No.	Dataset	Features			No. of classes	Train size	Test size
		All	Nominal	Continuous			
1	jEdit	9	1	8	2	369	CV
2	AR5	30	1	29	2	36	CV
3	MC1	39	1	38	2	9466	CV
4	CM1	22	1	21	2	498	CV
5	KC1_Mod	95	1	94	2	282	CV

feature vector and C is a class [58]. By independence, it is meant as probabilistic independence, that is, A is independent of B given C whenever $Pr(A|B, C) = Pr(A|C)$ for all possible values of A, B and C, whenever $Pr(C) > 0$ [57].

C4.5 is the most well-known algorithm in the literature for building decision trees [59, 48]. C4.5 first creates a decision-tree based on the attribute values of the available training data such that the internal nodes denote the different attributes, the branches correspond to value of a certain attribute and the leaf nodes correspond to the classification of the dependent variable. The decision tree is made recursively by identifying the attribute(s) that discriminates the various instances most clearly, i.e., having the highest information gain. Once a decision tree is made, the prediction for a new instance is done by checking the respective attributes and their values.

We have applied the selected FSS methods to five real-world datasets from the PROMISE repository [26]. These data sets are jEdit, AR5, MC1, CM1 and KC1_Mod. The datasets are available in ARFF (Attribute-Relation File Format), useable in the open source machine learning tool called WEKA (Waikato Environment for Knowledge Analysis) [60]. The datasets are selected based on their variance in terms of number of instances and the number of attributes. The number of instances vary from being less than 50 up to several thousands, with the number of attributes varying from being in a single digit to nearly a hundred. The characteristics of datasets are given in Table 2.

The source of jEdit data set is jEdit editor source code in Java and its Apache Subversion (SVN) log data. The data set contains metrics data computed by Understand C++ metric tool while bug data is extracted from SVN log files. The metrics are computed for jEdit release 4.0 while bugs are calculated between the releases 4.0 and 4.2. The source of AR5 data set is an embedded software used in manufacturing and implemented in C. Function/method level static code attributes are collected using Prest Metrics Extraction and Analysis Tool. The rest of the data sets (MC1, CM1, KC1_Mod) are NASA Metrics Data Program defect data sets. The metrics data consist of static code measures such as The McCabe and Halstead measures.

We restrict ourselves to evaluate the performance of *binary* classifiers which categorizes instances or software modules as being either fault-prone (*fp*) or non-fault prone (*nfp*). We are interested in predicting whether or not a module contains any faults, rather than the total number of faults. A common assessment

Module actually has faults

		NO	YES
Classifier predicts faults	NO	TN: True Negative	FN: False Negative
	YES	FP: False Positive	TP: True Positive

Fig. 3 The fault prediction sheet (confusion matrix)

procedure for binary classifiers is to count the number of correctly predicted modules over hold-out (test set) data. A fault prediction sheet [61]. as in Fig. 3, is commonly used.

Based on the different possibilities in the fault prediction sheet, various measures are typically derived. El-Emam et al. [62] have derived a number of measures based on this; the most common ones being rate of faulty module detection (or probability of detection (*PD*) or specificity), overall prediction accuracy (*acc*), probability of false alarm (*PF* or recall) and precision (*prec*). However the measure of overall accuracy *acc* has been criticized as being misleading since it ignores the data distribution and cost information [63]. The other measures of *PD, PF* and *prec* also reveal only one aspect of the prediction models at a time; thus their use introduces bias in performance assessment. Use of these measures also complicate comparisons and model selection since there is always a tradeoff between three measures, e.g. one model might exhibit a high *PD* but lower *prec* [63].

A receiver operating characteristic (ROC) curve [64] and the area under a ROC curve (AUC) [65] have been shown to be more statistically consistent and discriminating than predictive accuracy, *acc* [66]. The ROC curve is also a more general way, than numerical indices, to measure a classifier's performance [67]. A ROC curve provides an intuitive way to compare the classification performances of different techniques. ROC is a plot of the trade-off between the ability of the classifier to correctly detect fault-prone modules (*PD*) and the number of non-fault prone modules that are incorrectly classified (*PF*) across all possible experimental threshold settings [63, 68]. In short the (*PF, PD*) pairs generated by adjusting the algorithms threshold settings forms an ROC curve. A typical ROC curve is shown in Fig. 4.

This concave curve has the probability of detection (*PD*) on *y*-axis while the *x*-axis shows the probability of false alarms (*PF*). The start and end points for the ROC curve are (0, 0) to (1, 1), respectively. The software engineers need to identify the points on the ROC curve that suits their risks and budgets for the project [69]. A straight line from (0, 0) to (1, 1) offers no information while the point (*PF* = 0, *PD* = 1) is the ideal point on the ROC curve. A negative curve bends away from the ideal point while a preferred curve bends up towards the ideal point. As such, if we can divide the ROC space into four regions as shown in Fig. 5, the only region with practical value for software engineers is region *A* with acceptable *PD* and *PF* values. The regions *B, C* and *D* represent poor classification performance and hence are of little to no interest to software engineers [63].

Fig. 4 A typical ROC curve

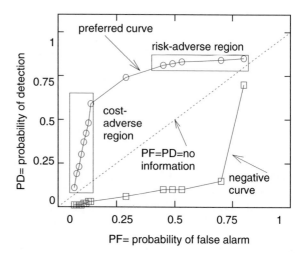

Fig. 5 Four regions in the ROC space

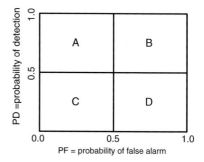

Area under the curve (AUC) [70] acts as a single scalar measure of expected performance and is an obvious choice for performance assessment when ROC curves for different classifiers intersect [4] or if the algorithm does not allow configuring different values of the threshold parameter. AUC, as with the ROC curve, is also a general measure of predictive performance since it separates predictive performance from class and cost distributions [4]. The AUC measures the probability that a randomly chosen *fp* module has a higher output value than a randomly chosen *nfp* module [64]. The value of AUC is always between 0 and 1; with a higher AUC indicating that the classifier is on average more to the upper left region *A* in Fig. 5.

We have used AUC as a measure of classification performance for the different FSS methods. For all the datasets, the AUC value averaged over 10 fold cross-validation runs, was calculated for each FSS method-dataset combination before and after FSS. For each cross-validation fold, the FSS method reduced the number of features in the dataset before being passed to C4.5 and NB classifiers.

5 Results and Analysis

Table 3 show results for all the datasets for FSS with NB. This table shows the AUC statistic for each FSS method and along with the AUC statistic when no feature selection is performed (the second column). The values in bold indicate if the use of the FSS method leads to an improvement of the AUC value, in comparison with when no FSS method is used. A number of FSS methods give an improved AUC value in comparison with the original AUC value without any feature selection. However, we need to test for any statistically significant differences between the different groups of AUC values. Since we have more than two samples with non-normal distributions, the Kruskal-Wallis test with significance level of 0.05 is used to test the null hypothesis that all samples are drawn from the same population. The result of the test ($p = 0.86$) suggested that it is not possible to reject the null hypothesis and, thus, there is no difference between any of the AUC values for the different FSS methods using NB **and** the AUC values of using NB as a classifier before and after applying the FSS methods.

Table 4 shows the number of attributes selected by each FSS method for NB. Wrapper, CFS, Relief and GP produce comparable AUC values with fewer number of selected features. PCA and IG, on the other hand, tend to select a much wider range of features to provide comparable classification results using NB.

Table 5 shows the AUC statistic for each FSS method using C4.5 along with the AUC statistic when no feature selection is used (second column). Again, the values in bold indicate that the use of FSS method leads to an improvement of the AUC value, in comparison with when no FSS is used. The result show that multiple FSS methods do improve the classification performance across all data sets. However, the result of using the Kruskal-Wallis test with $\alpha = 0.05$ ($p = 0.628$) suggested that it is not possible to reject the null hypothesis of all samples being drawn from the same population. Thus there is no significant difference between: (a) any of the AUC values for the different FSS methods using C4.5 and (b) the AUC values of using C4.5 as a classifier before and after applying the FSS methods.

Table 6 shows the number of attributes selected by each FSS method for C4.5. WRP, IG, CFS and GP produce comparable average AUC values with fewer number of selected features. RLF, PCA and CNS tend to select a wider range of features to provide comparable classification results using C4.5.

Table 3 FSS results with NB

Dataset	NB	IG	RLF	PCA	CFS	CNS	WRP	GP
jEdit	0.659	**0.67**	**0.67**	0.629	**0.668**	**0.67**	0.629	**0.67**
AR5	0.907	**0.933**	**0.942**	**0.938**	**0.942**	0.866	0.875	**0.915**
MC1	0.909	**0.919**	**0.92**	0.907	0.881	0.906	0.794	**0.93**
CM1	0.658	**0.718**	**0.728**	0.653	**0.691**	**0.685**	**0.738**	**0.68**
KC1_Mod	0.78	**0.851**	**0.938**	**0.854**	**0.84**	**0.86**	**0.802**	**0.87**

Table 4 Number of features selected by each FSS method for NB. The figures in % indicate the percentage of original features retained

Dataset	Org.	IG	RLF	PCA	CFS	CNS	WRP	GP
jEdit	9	6 (66.67 %)	6 (66.67 %)	5 (55.55 %)	5 (55.55 %)	6 (66.67 %)	3 (33.33 %)	2 (22.22 %)
AR5	30	4 (13.33 %)	2 (6.67 %)	7 (23.33 %)	2 (6.67 %)	2 (6.67 %)	1 (3.33 %)	6 (20 %)
MC1	39	20 (51.28 %)	12 (30.77 %)	14 (35.90 %)	4 (10.26 %)	15 (38.46 %)	1 (2.56 %)	4 (10.26 %)
CM1	22	5 (22.73 %)	3 (13.64 %)	4 (18.18 %)	7 (31.82 %)	11 (50 %)	2 (9.09 %)	14 (63.64 %)
KC1_Mod	95	3 (3.16 %)	8 (8.42 %)	17 (17.89 %)	8 (8.42 %)	2 (2.10 %)	4 (4.21 %)	7 (7.37 %)
Average	39	7.6 (19.49 %)	6.2 (15.90 %)	9.4 (24.10 %)	5.2 (13.33 %)	7.2 (18.46 %)	2.2 (5.64 %)	6.6 (16.92 %)

Table 5 FSS results with C4.5

Dataset	NB	IG	RLF	PCA	CFS	CNS	WRP	GP
jEdit	0.594	**0.644**	**0.623**	**0.636**	**0.612**	0.592	0.636	**0.62**
AR5	0.717	**0.817**	**0.866**	**0.763**	**0.866**	0.757	**0.817**	**0.797**
MC1	0.791	**0.829**	**0.796**	0.708	**0.795**	0.776	0.747	**0.854**
CM1	0.558	**0.615**	**0.587**	0.506	0.542	**0.596**	0.49	**0.644**
KC1_Mod	0.599	**0.806**	**0.684**	0.555	0.553	0.589	0.579	**0.69**

As is clear from the above discussion, NB and C4.5 show insignificantly different classification accuracies for the variety of FSS methods used. This result is in agreement with the study by Hall and Holmes [15] where the authors concluded that there is no single best approach for FSS for all situations. Song et al. [6] and Menzies et al. [38] also reach a similar conclusion:

> [...] we see that a data preprocessor/attribute selector can play different roles with different learning algorithms for different data sets and that no learning scheme dominates, i.e., always outperforms the others for all data sets. This means we should choose different learning schemes for different data sets, and consequently, the evaluation and decision process is important [6].

> [...] the best attribute subsets for defects predictors can change dramatically from data set to data set. Hence, conclusions regarding the best attribute(s) are very brittle, i.e., may not still apply when we change data sets [38].

Below we discuss the individual AUC values given for NB and C4.5 for different FSS methods.

From Table 3, it can be seen that for attribute selection with NB, the best AUC values are from three FSS methods (RLF, IG and GP) that improve NB on all five data sets and degrade it on none. CFS is the second best with improvement on four data sets and degradation on one. CNS, WRP and PCA give better performance on two data sets but also degrade performance on three data sets.

An overall pattern that is clear from Table 3 is that FSS is generally useful for NB's application to software fault prediction studies without significantly affecting classification accuracy. The results for NB in this study differ with the results given in the study by Hall and Holmes [15]. In that study, WRP was a clear winner in accuracy for NB. The potential reason for this performance could be attributed to the nature of the forward selection search in WRP which is used to generate the ranking such that strong attribute rankings are not identified. This search mechanism potentially works well in tandem with NB which has an attribute independence assumption [15]. However our results suggest that WRP is not at all a clear winner for NB where other FSS methods are also giving statistically insignificant results. This suggests that there are reasons other than the attribute independence assumption of NB that affects classification accuracy of NB with different FSS methods.

Table 6 Number of features selected by each FSS method for C4.5. The figures in % indicate the percentage of original features retained

Dataset	Org.	IG	RLF	PCA	CFS	CNS	WRP	GP
jEdit	9	3 (33.33 %)	4 (44.44 %)	5 (55.55 %)	5 (55.55 %)	6 (66.67 %)	5 (55.55 %)	2 (22.22 %)
AR5	30	1 (3.33 %)	2 (6.67 %)	7 (23.33 %)	2 (6.67 %)	2 (6.67 %)	1 (3.33 %)	6 (20 %)
MC1	39	9 (23.08 %)	19 (48.72 %)	14 (35.90 %)	4 (10.26 %)	15 (38.46 %)	1 (2.56 %)	4 (10.26 %)
CM1	22	2 (9.09 %)	9 (40.91 %)	4 (18.18 %)	7 (31.82 %)	11 (50 %)	2 (9.09 %)	14 (63.64 %)
KC1_Mod	95	3 (3.16 %)	7 (7.37 %)	17 (17.89 %)	8 (8.42 %)	2 (2.10 %)	4 (4.21 %)	7 (7.37 %)
Average	39	3.6 (9.23 %)	8.2 (21.02 %)	9.4 (24.10 %)	5.2 (13.33 %)	7.2 (18.46 %)	2.6 (6.67 %)	6.6 (16.92 %)

In terms of number of features selected for NB, the methods retaining the least number of features on average are WRP, CFS, RLF and GP. From Table 4 it can be seen that CFS chooses fewer features to all other FSS methods. From the techniques that were better on accuracy based on AUC values for NB, i.e., RLF and GP, are also among the methods that retains the least number of features. This is encouraging and shows that RLF and GP produce higher AUC values for NB while retaining minimum number of features on average, considering the data sets used in the experiment. PCA turns out to be worst in terms of retaining few features.

From Table 5, one can see the individual AUC values for attribute selection with C4.5. The results are in agreement with the results from NB. The best FSS methods for C4.5 are IG, RLF and GP which improve C4.5's performance on five data sets and degrade it on none. CFS improve C4.5's performance on three data sets and degrades it on two. CNS and PCA improve C4.5's performance on two data sets and degrades it on three. Result for WRP is that it degrades performance on four data sets and improves it on one.

As with NB, an overall pattern clear from Table 5 is that FSS is generally useful for C4.5's application to software fault prediction without significantly affecting classification accuracy. According to the study by Hall and Holmes [15]. *"The success of ReliefF and consistency with C4.5 could be attributable to their ability to identify attribute interactions (dependencies). Including strongly interacting attributes in a reduced subset increases the likelihood that C4.5 will discover and use interactions early on in tree construction before the data becomes too fragmented"*. The fact that we did not get consistent results with both RLF and CNF allows us to suggest that the ability to identify attribute interactions (dependencies) might not be the only differentiating factor in classification accuracy with respect to C4.5. As was our argument in case of NB, we argue that there are factors other than the ability to identify attribute interactions that are affecting classification accuracy of C4.5.

In terms of number of features retained for C4.5 (Table 6), WRP retains the minimum percentage of features on average, followed by IG, CFS, GP, CNS and RLF respectively. PCA is the worst in terms of retaining features for C4.5 with 24.10 %. Our results show WRP as a clear winner in our case while CFS is at third place in terms of retaining the minimum number of features on average. From the methods that were better on accuracy based on AUC values for C4.5 (IG, RLF and GP), IG and GP are at second and fourth place respectively in terms of retaining minimum percentage of features on average. This might suggest that IG and GP are suitable FSS methods for C4.5 considering the data sets we used in this study. RLF is down in ranking in Table 6, however its larger feature set sizes are justified by higher classification accuracy than the other methods.

Below we summarize the results of our study:

- FSS is useful and generally improves classification accuracy.
- There are no statistically significant differences for either NB or C4.5 for the variety of FSS methods used.
- Based on individual AUC values, IG, RLF and GP improve NB and C4.5 on five data sets and degrade them on none.

- CFS, RLF and GP retain the minimum percentage of features on average for NB.
- PCA is the worst in terms of retaining the minimum percentage of features on average for NB.
- There are factors other than the attribute independence assumption of NB that affect its classification accuracy with different FSS methods.
- IG, RLF and GP improve importance of C4.5 on five data sets and degrades it on none.
- WRP and CFS retains the minimum percentage of features on average for C4.5.
- There are factors other than the ability to identify attribute interactions that are affecting classification accuracy of C4.5.

After having discussed the results, we come to a crucial question: If various FSS methods perform differently for different machine learning algorithms, what factors are most important to consider while selecting FSS methods to use? Hall and Holmes [15] argue in their paper that there are three factors to consider:

1. An understanding of how different FSS methods work.
2. Strengths and weaknesses of the target learning algorithm.
3. Background knowledge about data.

While agreeing to all of the above factors, we add that if the goal is to improve classification accuracy of a learner, a decision about selecting a FSS method has to be reached in combination with following additional criteria:

1. Choice of resampling method.
2. Choice of data filtering technique (to address class imbalance, outlier removal, handling missing values and discretizing numeric attributes).
3. Choice of accuracy measure to use.

Choice of a resampling method concerns how to divide historical data into training and test data. In order to assess the generalizability of a learner, it is necessary that the test data are not used in anyway to build the learners [6]. A recent study by Afzal et al. [8] recommended the use of bootstrapping for software defect prediction studies. If not bootstrapping, the second recommended choice is to use leave-one-out cross validation for smaller data sets and 10-fold cross validation for large data sets. This subject however require more empirical studies to further strengthen these recommendations.

We also argue that the role of a data filtering technique in accurately classifying software components is important. A study by Gao et al. [44] demonstrated that data sampling (over-sampling or under-sampling) can counteract the adverse effect attributed to class imbalance in software fault prediction. They also concluded that feature selection became more efficient when used after data sampling. There are other examples of the use of data filtering techniques in software fault prediction, e.g., Menzies et al. [38] and Song et al. [6] used a log filtering preprocessor which replaces all numerics with their logarithms.

Choice of an accuracy indicator to evaluate the performance of defect predictors is also an important decision criterion. The use of MMRE as an accuracy indicator has been criticized by several authors [7, 71, 72]. Consequently, area under the receiver operating characteristic curve (AUC) is increasingly being used as a standard choice for performance evaluation in software fault predictions studies.[2]

It is important to highlight the performance of an evolutionary computational method (GP) as a FSS method. For both NB and C4.5, GP improved the AUC values for maximum number of data sets and degraded on the least number of data sets. For NB, GP is also among the methods that retained minimum percentage of features on average. It is worth noting that for GP feature selection is an implicit part of GP evolution. This enables automatic or semi-automatic selection of features during model generation. GP allows almost any combination of a number of features. Evolution can freely add remove multiple features and can reconsider previous selections as new combinations are tried [73]. A potential disadvantage of using an evolutionary algorithm like GP is that it can take more computational resources as compared with other methods. Therefore with GP it has to be a tradeoff between how much improvement in classification accuracy is required against available resources.

6 Validity Evaluation

Wohlin et al. [74] discuss four types of threats to an experimental study: external (ability to generalize), conclusion (ability to apply statistical tests), internal (ability to correctly infer connections between dependent and independent variables) and construct (ability of dependent variable to capture the effect being measured).

External validity The datasets used in this study represent real-world use, collected during the course of real industry projects developed by professionals. The datasets differed in their number of attributes and sizes. However as noted by Gao et al. [44] analysis of another data set from different application domain may provide different results which is a likely threat in all empirical software engineering research.

Conclusion validity We were mindful that the type of statistical tests could potentially affect end results, therefore Kruskal Wallis test was used as we had more than two samples with non-normal distributions. This empirical study was performed using 10-fold cross-validation for statistically reliable results (recommended in Kohavi [75] and Afzal et al. [8]). The performance of classifiers is compared using area under the receiver operating characteristic curve (AUC) which we motivate is a standard way of evaluating classification results.

Internal validity According to Gao et al. [44], different factors can affect the internal validity of fault proneness estimates: measurement errors while collecting

---

[2]Section 4 provides more details about AUC.

and recording software metrics; modeling errors due to the unskilled use of software applications; errors in model selection during the modeling process; and the presence of outliers and noise in the training dataset. We used the publicly available data sets so other researchers can replicate our work. Secondly we have given the parameter settings for different methods used to ease replication of our work.

Construct validity The datasets used in this study are the ones donated by the authors of fault prediction studies and mostly use structural measures. Structural measures are widely used in software fault prediction studies [39], however finding the right predictors for software fault proneness is an active area of research.

7 Conclusions

Feature subset selection (FSS) methods are used to keep the number of features in a dataset as small as possible. Out of the various perceived advantages of using these FSS methods (Sect. 1), this study evaluate whether or not the use of FSS methods have any significant affect on the classification accuracy of software fault prediction when used with two diverse learning algorithms, C4.5 and naïve Bayes.

We compare a total of seven FSS methods, representing a mix of state-of-the-art methods and an evolutionary computation method, on five software fault prediction datasets from the PROMISE data repository. Our findings show that feature subset selection is generally useful for software fault prediction using naïve Bayes and C4.5. However there are no clear winners for either of the two learning algorithms for the variety of FSS methods used.

Based on individual AUC values, IG, RLF and GP improve naïve Bayes and C4.5 on five data sets and degrade it on none. RLF, GP and CFS also retain the minimum percentage of features on average for naïve Bayes. WRP and CFS retain the minimum percentage of features on average for C4.5.

In summary, our results suggest that RLF, IG and GP are the best FSS methods for software fault prediction accuracy using naïve Bayes and C4.5. CNS and CFS are also good overall performers. We recommend that any future software fault prediction study be preceded by an initial analysis of FSS methods, not missing on methods that have shown to be more consistent than their competitors. It is recommended in literature that for selecting a FSS method, a data miner needs to have an understanding of how different FSS methods work, strengths and weaknesses of the target learning algorithm and background knowledge about data. In the context of software fault prediction studies, we additionally recommend that the data miner needs to have an understanding of different resampling methods, data filtering techniques and accuracy measures for increasing the reliability and validity of prediction results. In this study, we do not offer an interpretation of features retained by different FSS methods. It is, nevertheless, an interesting future work to relate features retained by different FSS methods with functioning of the target learning algorithm and background knowledge about data.

References

1. Khoshgoftaar, T.M., Seliya, N.: Fault prediction modeling for software quality estimation: Comparing commonly used techniques. Empirical Softw. Eng. **8**(3), 255–283 (2004)
2. Catal, C., Diri, B.: A systematic review of software fault prediction studies. Expert Syst. Appl. **36**(4), 7346–7354 (2009)
3. Hall, T., Beecham, S., Bowes, D., Gray, D., Counsell, S.: A systematic review of fault prediction performance in software engineering. IEEE Trans. Softw. Eng. (99) (2011)
4. Lessmann, S., Baesens, B., Mues, C., Pietsch, S.: Benchmarking classification models for software defect prediction: a proposed framework and novel findings. IEEE Trans. Softw. Eng. **34**(4), 485–496 (2008)
5. Fenton, N.E., Neil, M.: A critique of software defect prediction models. IEEE Trans. Softw. Eng. **25**(5), 675–689 (1999)
6. Song, Q., Jia, Z., Shepperd, M., Ying, S., Liu, J.: A general software defect-proneness prediction framework. IEEE Trans. Softw. Eng. **37**(3), 356–370 (2011)
7. Foss, T., Stensrud, E., Kitchenham, B.A., Myrtveit, I.: A simulation study of the model evaluation criterion MMRE. IEEE Trans. Softw. Eng. **29**(11) (2003)
8. Afzal, W., Torkar, R., Feldt, R.: Resampling methods in software quality classification. Int. J. Software Eng. Knowl. Eng. **22**, 203–223 (2012)
9. Gray, D., Bowes, D., Davey, N., Sun, Y., Christianson, B.: The misuse of the NASA metrics data program data sets for automated software defect prediction. IET Semin. Dig. **1**, 96–103 (2011)
10. Khoshgoftaar, T.M., Gao, K., Seliya, N.: Attribute selection and imbalanced data: Problems in software defect prediction. IEEE Computer Society, Los Alamitos, CA, USA (2010)
11. Shivaji, S., Whitehead, J.E.J, Akella, R., Kim, S. Reducing features to improve bug prediction. In: Proceedings of the 2009 IEEE/ACM International Conference on Automated Software Engineering (ASE'09), IEEE Computer Society, Washington, DC, USA (2009)
12. Rodriguez, D., Ruiz, R., Cuadrado-Gallego, J., Aguilar-Ruiz, J.: Detecting fault modules applying feature selection to classifiers. In: IEEE International Conference on Information Reuse and Integration (IRI'07) (2007a)
13. Rodriguez, D., Ruiz, R., Cuadrado-Gallego, J., Aguilar-Ruiz, J., Garre, M.: Attribute selection in software engineering datasets for detecting fault modules. In: 33rd EUROMICRO Conference on Software Engineering and Advanced Applications (EUROMICRO'07) (2007b)
14. Kohavi, R., John, G.H.: Wrappers for feature subset selection. Artif. Intell. **97**, 273–324 (1997)
15. Hall, M.A., Holmes, G.: Benchmarking attribute selection techniques for discrete class data mining. IEEE Trans. Knowl. Data Eng. **15**, 1437–1447 (2003)
16. Jain, A.K., Duin, R.P.W., Mao, J.: Statistical pattern recognition: a review. IEEE Trans. Pattern Anal. Mach. Intell. **22**, 4–37 (2000)
17. Chen, Z., Boehm, B., Menzies, T., Port, D.: Finding the right data for software cost modeling. IEEE Softw. **22**, 38–46 (2005)
18. Janecek, A., Gansterer, W., Demel, M., Ecker, G.: On the relationship between feature selection and classification accuracy. In: Proceedings of the 3rd Workshop on New Challenges for Feature Selection in Data Mining and Knowledge Discovery (FSDM'08), Microtome Publishing, Brookline, MA, USA (2008)
19. Burke, E.K., Kendall, G. (eds.): Search methodologies—Introductory tutorials in optimization and decision support techniques. Springer Science and Business Media, Inc., 233 Spring Street, New York, USA (2005)
20. Dybå, T., Kampenes, V.B., Sjøberg, D.I.: A systematic review of statistical power in software engineering experiments. Inf. Softw. Technol. **48**(8), 745–755 (2006)

21. Afzal, W., Torkar, R., Feldt, R., Gorschek, T.: Genetic programming for cross-release fault count predictions in large and complex software projects. In: Chis, M. (ed.) Evolutionary Computation and Optimization Algorithms in Software Engineering: Applications and Techniques, pp. 94–126. IGI Global, Hershey, USA (2009)
22. Muni, D., Pal, N., Das, J.: Genetic programming for simultaneous feature selection and classifier design. IEEE Trans. Syst. Man Cybern. B Cybern. **36**(1), 106–117 (2006)
23. Smith, M.G., Bull. L.: Feature construction and selection using genetic programming and a genetic algorithm. In: Proceedings of the 6th European Conference on Genetic Programming (EuroGP'03), Springer-Verlag, Berlin, Heidelberg (2003)
24. Vivanco, R., Kamei, Y., Monden, A., Matsumoto, K., Jin, D.: Using search-based metric selection and oversampling to predict fault prone modules. In: 2010 23rd Canadian Conference on Electrical and Computer Engineering (CCECE'10) (2010)
25. Yang, J., Honavar, V.: Feature subset selection using a genetic algorithm. IEEE Intell. Syst. and Their Appl. **13**(2), 44–49 (1998)
26. Boetticher, G., Menzies, T., Ostrand, T.: PROMISE repository of empirical software engineering data. http://promisedata.org/ repository, West Virginia University, Department of Computer Science (2007)
27. Molina, L.C., Belanche, L., Nebot, Àngela: Feature selection algorithms: a survey and experimental evaluation. Proceedings of the 2002 IEEE International Conference on Data Mining (ICDM'02), pp. 306–313. IEEE Computer Society, Washington, DC, USA (2002)
28. Guyon, I., Elisseeff, A.: An introduction to variable and feature selection. J. Mach. Learn. Res. **3**, 1157–1182 (2003)
29. Blum, A.L., Langley, P.: Selection of relevant features and examples in machine learning. Artif. Intell. **97**, 245–271 (1997)
30. Dash, M., Liu, H.: Feature selection for classification. Intelligent Data Analysis **1**(1–4), 131–156 (1997)
31. Liu, H., Yu, L.: Toward integrating feature selection algorithms for classification and clustering. IEEE Trans. Knowl. Data Eng. **17**(4), 491–502 (2005)
32. Dejaeger, K., Verbeke, W., Martens, D., Baesens, B.: Data mining techniques for software effort estimation: a comparative study. IEEE Trans. Softw. Eng. **38**, 375–397 (2012)
33. Chen, Z., Menzies, T., Port, D., Boehm, B.: Feature subset selection can improve software cost estimation accuracy. SIGSOFT Softw. Eng. Notes **30**(4), 1–6 (2005)
34. Menzies, T., Jalali, O., Hihn, J., Baker, D., Lum, K.: Stable rankings for different effort models. Autom. Softw. Eng. **17**, 409–437 (2010)
35. Kirsopp, C., Shepperd, M.J., Hart, J.: Search heuristics, case-based reasoning and software project effort prediction. Proceedings of the 2002 Genetic and Evolutionary Computation Conference (GECCO'02), pp. 1367–1374. Morgan Kaufmann Publishers Inc., San Francisco, CA, USA (2002)
36. Azzeh, M., Neagu, D., Cowling, P.: Improving analogy software effort estimation using fuzzy feature subset selection algorithm. In: Proceedings of the 4th International Workshop on Predictor Models in Software Engineering (PROMISE'08), ACM, New York, NY, USA (2008)
37. Li, Y., Xie, M., Goh, T.: A study of mutual information based feature selection for case based reasoning in software cost estimation. Expert Systems with Applications 36(3, Part 2):5921–5931 (2009)
38. Menzies, T., Greenwald, J., Frank, A.: Data mining static code attributes to learn defect predictors. IEEE Trans. Softw. Eng. **33**(1), 2–13 (2007)
39. Catal, C., Diri, B.: Investigating the effect of dataset size, metrics sets, and feature selection techniques on software fault prediction problem. Inf. Sci. **179**, 1040–1058 (2009)
40. Khoshgoftaar, T.M., Seliya, N., Sundaresh, N.: An empirical study of predicting software faults with case-based reasoning. Softw. Qual. Control **14**, 85–111 (2006)
41. Wang, H., Khoshgoftaar, T., Gao, K., Seliya, N.: High-dimensional software engineering data and feature selection. In: 21st International Conference on Tools with Artificial Intelligence (ICTAI'09), pp. 83–90 (2009)

42. Khoshgoftaar, T.M., Nguyen, L., Gao, K., Rajeevalochanam, J.: Application of an attribute selection method to CBR-based software quality classification. In: Proceedings of the 15th IEEE International Conference on Tools with Artificial Intelligence (ICTAI'03), IEEE Computer Society, Washington, DC, USA (2003)
43. Altidor, W., Khoshgoftaar, T.M., Gao, K.: Wrapper-based feature ranking techniques for determining relevance of software engineering metrics. Int. J. Reliab. Qual. Saf. Eng. 17, 425–464 (2010)
44. Gao, K., Khoshgoftaar, T., Seliya, N.: Predicting high-risk program modules by selecting the right software measurements. Softw. Qual. J. 20, 3–42 (2012)
45. Gao, K., Khoshgoftaar, T.M., Wang, H., Seliya, N.: Choosing software metrics for defect prediction: an investigation on feature selection techniques. Softw. Pract. Experience 41(5), 579–606 (2011)
46. Khoshgoftaar, T.M., Gao, K., Napolitano, A.: An empirical study of feature ranking techniques for software quality prediction. Int. J. Softw. Eng. Knowl. Eng. (IJSEKE) 22, 161–183 (2012)
47. Wang, H., Khoshgoftaar, T.M., Napolitano, A.: Software measurement data reduction using ensemble techniques. Neurocomputing 92, 124–132 (2012)
48. Quinlan, J.R.: C4.5: programs for machine learning. Morgan Kaufmann Publishers Inc., San Francisco, CA, USA (1993)
49. Novakovic, J.: Using information gain attribute evaluation to classify sonar targets. In: Proceedings of the 17th Telecommunications forum (TELFOR'09) (2009)
50. Kira, K., Rendell, L.A.: The feature selection problem: traditional methods and a new algorithm. In: Proceedings of the 10th National Conference on Artificial Intelligence (AAAI'92) (1992)
51. Sikonja, M., Kononenko, I.: An adaptation of relief for attribute estimation in regression. In: Proceedings of the 14th International Conference on Machine Learning (ICML'97) (1997)
52. Hall, M.A.: Correlation-based feature selection for discrete and numeric class machine learning. In: Proceedings of the 2000 International Conference on Machine Learning (ICML'00), Morgan Kaufmann Publishers Inc., San Francisco, CA, USA (2000)
53. Liu, H., Setiono, R.: A probabilistic approach to feature selection—A filter solution. Proceedings of the 1996 International Conference on Machine Learning (ICML'96), pp. 319–327. Morgan Kaufmann Publishers Inc., San Francisco, CA, USA (1996)
54. Poli, R., Langdon, W.B., McPhee, N.F.: A field guide to genetic programming. Published via http://lulu.com and freely available at http://www.gp-field-guide.org.uk. URL: http://www.gp-field-guide.org.uk, (with contributions by Koza, J.R.) (2008)
55. Koza, J.R.: Genetic programming: on the programming of computers by means of natural selection. MIT Press, Cambridge, MA, USA (1992)
56. Silva, S.: GPLAB—A genetic programming toolbox for MATLAB. http://gplab.sourceforge. net, Last checked: 22 Dec 2014 (2007)
57. Friedman, N., Geiger, D., Goldszmidt, M.: Bayesian network classifiers. Mach. Learn. 29(2–3), 131–163 (1997)
58. Rish, I.: An empirical study of the naive Bayes classifier. In: Proceedings of the workshop on empirical methods in AI (IJCAI'01) (2001)
59. Kotsiantis, S., Zaharakis, I., Pintelas, P.: Machine learning: a review of classification and combining techniques. Artif. Intell. Rev. 26(3), 159–190 (2007)
60. Hall, M., Frank, E., Holmes, G., Pfahringer, B., Reutemann, P., Witten, I.H.: The WEKA data mining software: an update. SIGKDD Explor. Newsl. 11, 10–18 (2009)
61. Menzies, T., DiStefano, J., Orrego, A., Chapman, R.M.: Assessing predictors of software defects. In: Proceedings of the Workshop on Predictive Software Models, collocated with ICSM'04. URL: http://menzies.us/pdf/04psm.pdf (2004)
62. El-Emam, K., Benlarbi, S., Goel, N., Rai, S.N.: Comparing case-based reasoning classifiers for predicting high risk software components. J. Syst. Softw. 55(3), 301–320 (2001)

63. Ma, Y., Cukic, B.: Adequate and precise evaluation of quality models in software engineering studies. In: Proceedings of the 3rd International Workshop on Predictor Models in Software Engineering (PROMISE'07), IEEE Computer Society, pp 1, Washington, DC, USA(2007)
64. Fawcett, T.: An introduction to ROC analysis. Pattern Recogn. Lett. **27**(8), 861–874 (2006)
65. Hanley, J.A., McNeil, B.J.: The meaning and use of the area under a receiver operating characteristic (ROC) curve. Radiology 143(1):29–36 (1982)
66. Ling, C.X., Huang, J., Zhang, H.: AUC: a statistically consistent and more discriminating measure than accuracy. In: Proceedings of the Eighteenth International Joint Conference on Artificial Intelligence (IJCAI'03) (2003)
67. Yousef, W.A., Wagner, R.F., Loew, M.H.: Comparison of non-parametric methods for assessing classifier performance in terms of ROC parameters. In: Proceedings of the 33rd Applied Imagery Pattern Recognition Workshop (AIPR'04), IEEE Computer Society, Washington, DC, USA (2004)
68. Jiang, Y., Cukic, B., Menzies, T., Bartlow, N.: Comparing design and code metrics for software quality prediction. In: Proceedings of the 4th international workshop on predictor models in software engineering (PROMISE'08), ACM, New York, NY, USA (2008)
69. Jiang, Y., Cukic, B., Menzies, T.: Fault prediction using early lifecycle data. In: Proceedings of the 18th IEEE International Symposium on Software Reliability (ISSRE'07), IEEE Computer Society, Washington, DC, USA (2007)
70. Bradley, A.P.: The use of the area under the ROC curve in the evaluation of machine learning algorithms. Pattern Recogn. **30**, 1145–1159 (1997)
71. Kitchenham, B.A., Pickard, L.M., MacDonell, S., Shepperd, M.: What accuracy statistics really measure? IEE Proc. Softw. **148**(3) (2001)
72. Myrtveit, I., Stensrud, E., Shepperd, M.: Reliability and validity in comparative studies of software prediction models. IEEE Trans. Softw. Eng. **31**(5), 380–391 (2005)
73. Langdon, W.B., Buxton, B.F.: Genetic programming for mining DNA chip data from cancer patients. Genet. Program Evolvable Mach. **5**, 251–257 (2004)
74. Wohlin, C., Runeson, P., Höst, M., Ohlsson, M., Regnell, B., Wesslén, A.: Experimentation in software engineering: an introduction. Kluwer Academic Publishers, USA (2000)
75. Kohavi, R.: A study of cross-validation and bootstrap for accuracy estimation and model selection. In: Proceedings of the 14th International Joint conference on Artificial Intelligence (IJCAI'95), Morgan Kaufmann Publishers Inc., San Francisco, CA, USA (1995)

Author Biographies

Wasif Afzal is a postdoctoral research fellow at Mälardalen University, Sweden. He received his PhD in Software Engineering from Blekinge Institute of Technology in 2011. His research interests are within software testing, prediction and estimation in software engineering and application of artificial intelligence techniques to software engineering problems.

Richard Torkar works as a professor of software engineering at Chalmers and the University of Gothenburg, Sweden. His interests lie mainly in quantitative software engineering and statistics.

Evolutionary Computation for Software Product Line Testing: An Overview and Open Challenges

Roberto E. Lopez-Herrejon, Javier Ferrer, Francisco Chicano, Alexander Egyed and Enrique Alba

Abstract Because of economical, technological and marketing reasons today's software systems are more frequently being built as families where each product variant implements a different combination of features. Software families are commonly called Software Product Lines (SPLs) and over the past three decades have been the subject of extensive research and application. Among the benefits of SPLs are: increased software reuse, faster and easier product customization, and reduced time to market. However, testing SPLs is specially challenging as the number of product variants is usually large making it infeasible to test every single variant. In recent years there has been an increasing interest in applying evolutionary computation techniques for SPL testing. In this chapter, we provide a concise overview of the state of the art and practice in SPL testing with evolutionary techniques as well as to highlight open questions and areas for future research.

Keywords Software product lines · Product line testing · Search based software engineering · Feature models · Feature set · Reverse engineering · Variability modeling

R.E. Lopez-Herrejon (✉) · A. Egyed
Software Systems Engineering Institute, Johannes Kepler University, Linz, Austria
e-mail: roberto.lopez@jku.at

A. Egyed
e-mail: alexander.egyed@jku.at

J. Ferrer · F. Chicano · E. Alba
Universidad de Málaga, Andalucía Tech, Sevilla, Spain
e-mail: ferrer@lcc.uma.es

F. Chicano
e-mail: chicano@lcc.uma.es

E. Alba
e-mail: eat@lcc.uma.es

© Springer International Publishing Switzerland 2016   59
W. Pedrycz et al. (eds.), *Computational Intelligence and Quantitative
Software Engineering*, Studies in Computational Intelligence 617,
DOI 10.1007/978-3-319-25964-2_4

1 Introduction

A *Software Product Line* (*SPL*) is a family of related software systems each of which provides a different combination of features [1], where a *feature* is commonly defined as an increment in program functionality [2]. Extensive research and practice attest to the substantial benefits of SPL practices such as increased software reuse, faster product customization, and reduced time to market (e.g. [3]). SPLs typically involve large number of software systems, which make it infeasible to individually test each one of them. To address this need, several testing techniques and approaches have been proposed, all with distinct advantages and drawbacks [4–7].

Search Based Software Engineering (*SBSE*) is an emerging discipline that focuses on the application of search-based optimization techniques to software engineering problems [8]. Among the techniques SBSE relies on is *evolutionary computation*—an area of artificial intelligence that studies algorithms that follow Darwinian principles of evolution [9]. Evolutionary computation techniques are generic, robust, and have been shown to scale to large search spaces. These properties have been extensively exploited for testing standard one-off systems (e.g. [10]), but their application to SPLs remains largely unexplored.

In this book chapter we present a concise overview of current techniques for SPL testing, describe and illustrate the salient work on evolutionary computation techniques applied to SPL testing, and highlight some of the open challenges that remain to be addressed. The chapter is structured as follows. Section 2 provides the basic background on SPL and evolutionary algorithms needed for this chapter. Section 3 provides a general overview on the state of the art of SPL testing. Section 4 describes *Combinatorial Interaction Testing* (*CIT*), the main approach for evolutionary SPL testing, and presents a simple illustrative algorithm that follows this approach. Section 5 presents a formal description of SPL testing as a multi-objective optimization problem, describes an algorithm to compute exact Pareto fronts, and summarizes the state of the art of research in this area. Section 7 summarizes the open questions and challenges. Section 8 presents the conclusions to our work.

2 Background

In this section we provide the basic background on the two topics that crosscut the chapter: Software Product Lines and Evolutionary Algorithms.

2.1 SPL Foundations—Feature Models and Running Example

Feature models have become a de facto standard for modelling the common and variable features of an SPL [11]. Features are depicted as labelled boxes and their relationships as lines, collectively forming a tree-like structure. Feature models then denote the set of feature combinations that the systems of an SPL can have [11, 12].

Figure 1 shows the feature model of our running example, the *Graph Product Line (GPL)*, a standard SPL of basic graph algorithms that has been extensively used as a case study in the SPL community [13]. In this SPL, a software system has feature GPL (the root of the feature model) which contains its core functionality, and a driver program (Driver) that sets up the graph examples (Benchmark) to which a combination of graph algorithms (Algorithms) are applied. The graphs (GraphType) can be either directed (Directed) or undirected (Undirected), and can optionally have weights (Weight). Two graph traversal algorithms (Search) can be optionally provided: Depth First Search (DFS) or Breadth First Search (BFS). A software system must provide at least one of the following algorithms: numbering of nodes in the traversal order (Num), connected components (CC), strongly connected components (SCC), cycle checking (Cycle), shortest path (Shortest), minimum spanning trees with Prim's algorithm (Prim) or Kruskal's algorithm (Kruskal).

In a feature model, each feature has exactly one parent feature and can have a set of child features. A child feature can only be selected in a feature combination of a valid software system if its parent is selected as well. The exception is the root feature that does not have any parent and it is always selected in any software system of a SPL. There are four kinds of feature relationships:

– *Mandatory features* are depicted with a filled circle. A mandatory feature is selected whenever its respective parent feature is selected. For example, features Algorithms and GraphType.

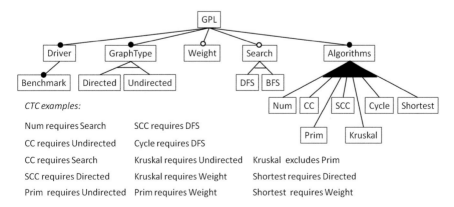

Fig. 1 Graph Product Line Feature Model [13]

- *Optional features* are depicted with an empty circle. An optional feature may or may not be selected if its respective parent feature is selected. An example is feature `Search`.
- *Exclusive-or relations* are depicted as empty arcs crossing over the lines connecting a parent feature with its child features. They indicate that *exactly one* of the features in the exclusive-or group must be selected whenever the parent feature is selected. For example, if feature `GraphType` is selected, then either feature `Directed` or feature `Undirected` must be selected.
- *Inclusive-or relations* are depicted as filled arcs crossing over a set of lines connecting a parent feature with its child features. They indicate that *at least one* of the features in the inclusive-or group must be selected if the parent is selected. If for instance, feature `Algorithms` is selected then at least one of the features `Num`, `CC`, `SCC`, `Cycle`, `Shortest`, `Prim`, or `Kruskal` must be selected.

Besides the parent-child relations, features can also relate across different branches of the feature model with *Cross-Tree Constraints (CTCs)*. Figure 1 textually shows the CTCs of GPL. For instance, `Cycle requires DFS` means that whenever feature `Cycle` is selected, feature `DFS` must also be selected. As another example, `Prim excludes Kruskal` means that both features cannot be selected at the same time in any product. These constraints as well as those implied by the hierarchical relations between features are usually expressed and checked using propositional logic, for further details refer to [12]. Now we present the basic definitions on which SPL testing terminology is defined in the next section.

Definition 1 (*Feature list*) A feature list (FL) is the list of features in a feature model.

The `FL` for the GPL feature model is [`GPL`, `Driver`, `Benchmark`, `GraphType`, `Directed`, `Undirected`, `Weight`, `Search`, `DFS`, `BFS`, `Algorithms`, `Num`, `CC`, `SCC`, `Cycle`, `Shortest`, `Prim`, `Kruskal`].

Definition 2 (*Feature set*) A feature set `fs` is a 2-tuple [`sel`,$\overline{\text{sel}}$] where `fs.sel` and `fs.`$\overline{\text{sel}}$ are respectively the set of selected and not-selected features in a system part of a SPL. Let `FL` be a feature list, thus `sel`,$\overline{\text{sel}}\subseteq$`FL`, `sel`$\cap\overline{\text{sel}} = \varnothing$, and `sel`$\cup\overline{\text{sel}} = FL$. Wherever unambiguous we use the term **product** as a synonym of feature set.

Definition 3 (*Valid feature set*) A feature set `fs` is valid with respect to a feature model *fm* iff `fs.sel` and `fs.`$\overline{\text{sel}}$ do not violate any constraints described by *fm*. The set of all valid feature sets represented by *fm* is denoted as $\mathcal{FS}^{fm}$.

GPL has 73 distinct valid feature sets, some of them depicted in Table 1, where selected features are ticked (✓) and unselected features are empty. An example of valid feature set is `fs1` that computes the algorithms `Kruskal` and `CC`, on `Undirected` graphs using `DFS` search. Thus, the selected features are `fs1.sel`={`GPL`, `Driver`, `GraphType`, `Weight`, `Search`, `Algorithms`, `Benchmark`, `Undirected`, `DFS`, `CC`, `Kruskal`}, and the unselected features `fs1.`$\overline{\text{sel}}$={`Directed`, `BFS`, `Num`, `SCC`, `Cycle`, `Shortest`,

Table 1 Sample feature sets of GPL

FS	GPL	Dri	Gtp	W	Se	Alg	B	D	U	DFS	BFS	N	CC	SCC	Cyc	Sh	Prim	Kru
fs0	✓	✓	✓	✓		✓	✓		✓								✓	
fs1	✓	✓	✓	✓	✓	✓	✓		✓	✓			✓					✓
fs2	✓	✓	✓		✓	✓	✓	✓		✓		✓			✓			
fs3	✓	✓	✓	✓	✓	✓	✓	✓			✓	✓				✓		
fs4	✓	✓	✓	✓	✓	✓	✓	✓		✓		✓		✓	✓	✓		
fs5	✓	✓	✓	✓	✓	✓	✓		✓	✓		✓	✓		✓			
fs6	✓	✓	✓	✓	✓	✓	✓		✓		✓	✓	✓				✓	
fs7	✓	✓	✓	✓	✓	✓	✓		✓	✓		✓	✓		✓			✓

Driver (Dri), GraphType (Gtp), Weight (W), Search (Se), Algorithms (Alg), Benchmark (B), Directed (D), Undirected (U), Num (N), Cycle (Cyc), Shortest (Sh), Kruskal (Kr)

`Prim`}. Consider now another feature set gs with selected features `BFS` and `Cycle`, meaning {`BFS,Cycle`} ⊂ `gs.sel`. This feature set is invalid because these two features violate the CTC that establishes that whenever `Cycle` feature is selected then feature `DFS` must be selected, i.e. `Cycle requires DFS`.

2.2 Basics of Evolutionary Algorithms

Evolutionary Computation is an area of computer science, artificial intelligence more concretely, that studies algorithms that follow Darwinian principles of evolution [9]. Algorithm 1 sketches the general structure of an evolutionary algorithm adapted from [9, 14]. It starts by creating an initial population of candidate solutions for the particular problem to address (Lines 1–2). The population is denoted by term $P(t)$ where t stands for the generation of the population. A measure of fitness to solve the problem is used to evaluate each member of the population (Line 3). Then, while not reaching a termination condition such as a given number of generations or fitness threshold (Lines 4–9), a new population is selected from the previous population and the newly created offspring (Lines 5–6). The new population is randomly mutated to promote solution diversity (Line 7) and is subsequently re-evaluated (Line 8).

Algorithm 1. Basic Evolutionary Algorithm

1: $t \leftarrow 0$
2: *initialize* $P(t)$
3: *evaluate* $P(t)$
4: **while not** *terminationCondition* **do**
5:     $t \leftarrow t + 1$
6:     *select* $P(t)$ *from* $P(t - 1)$
7:     *mutate* $P(t)$
8:     *evaluate* $P(t)$
9: **end while**

There are several types of evolutionary algorithms [9]; however, *Genetic Algorithms (GA)* are undoubtedly the most commonly used ones [15]. They typically employ a binary list representation and are commonly used for optimization problems such as job scheduling problems. In the coming sections we will explain how this basic algorithm is adapted for the problem of SPL testing.

3   Overview of SPL Testing

As SPL development practices become more prevalent, there is an increasing need of adequate and scalable SPL testing techniques. In recent years, there has been a growing interest by the research and practitioners communities to propose and evaluate new methods and tools to address this need. The results have been captured and analyzed in several systematic mapping studies and surveys. In this section, we summarize the most salient works among such studies and surveys to provide the context on which to place evolutionary computation techniques for SPL.

Engström and Runerson [4] performed a mapping study that takes a higher level view of the subject by focusing for instance on the organization and process for testing, and the type of testing techniques performed such as acceptance testing, unit testing, or integration testing. A similar and complementary mapping study was carried out by Neto et al. [5]. They analyzed, for instance, the different strategies that have been taken for SPL testing such as testing product by product, incremental testing (i.e. first product tested individually and the following products with regression testing), or opportunistic reuse (i.e. employ test assets available from other products). They also analyzed others factors and aspects of SPL testing such as the use of static and dynamic analysis techniques, the effort reduction, or non-functional testing. Among their most salient findings, is the lack of evidence on when to select a given testing strategy considering factors such as development processes employed or delivery time and budget constraints. The complementary nature of these two studies has been further analyzed and reported [16, 17].

More recent studies by do Carmo Machado et al. [7, 18] have taken a closer look at the techniques used for SPL testing. They classify the works in two so-called *interests*: papers that focus on selecting the products of the SPL to test, and papers that describe approaches to actually carry out the testing on the selected products. Their study found that the *de facto* approach for selecting which SPL products to test is *Combinatorial Interaction Testing* (*CIT*) which aims at constructing *samples* to drive the systematic testing of software system configurations [19, 20]. For the second interest, they found an array of techniques, mostly based on extensions to UML activity and sequence diagrams.[1] Their study highlights also several shortcomings, such as the lack of robust empirical evaluation and adequate tool support.

We should point out that CIT is a generic testing approach not only applicable in the context of SPL testing. In this general sense, CIT consists of four phases [21]: (i) *modeling* whose goal is to model the *System Under Test* (*SUT*) and its input space, (ii) *sampling* which produces a set of configurations that will be used for testing, (iii) *testing* that actually carries of the test based on different CIT parameters, and (iv) *analysis* where the results obtained are examined to identify faults and their underlying causes. In other words, the first two stages deal with the *what* should be tested, whereas the last two stages deal with the *how* should be tested [21].

Within the area of Search-Based Software Engineering a major research focus has been software testing [8, 22]. A recent overview by McMinn [10] highlights the major achievements made in the area and some of the open questions and challenges. We have performed a systematic mapping study whose focus is on SBSE

---

[1]http://www.uml.org/.

techniques applied to SPLs [23],[2] among such techniques are those based on evolutionary computation. Overall we found that almost all the research on evolutionary computation applied to SPL testing falls within the first two stages of CIT, modeling (based on feature models) and sampling (using different techniques), as we elaborate more on next section.

4 Combinatorial Interaction Testing for Software Product Lines

When Combinational Interaction Testing is applied to SPLs, the goal is to select a representative subset of products where interaction errors are more likely to occur rather than testing the complete product family [19]. In this section, we provide the basic terminology of CIT for SPLs,[3] use a simple evolutionary algorithm to illustrate CIT for the case of pairwise testing, and presents an overview of state of the art in CIT for SPL testing. In Sect. 5, we address the case when optimization of multiple objectives is considered.

4.1 Basic Terminology

Definition 4 (*t-set*) A t-set ts is a 2-tuple [sel, $\overline{\text{sel}}$] representing a partially configured product, defining the selection of t features of the feature list *FL*, i.e. ts.sel∪ts.$\overline{\text{sel}}$⊆FL ∧ ts.sel∩ts.$\overline{\text{sel}}$ = ∅ ∧ |ts.sel∪ts.$\overline{\text{sel}}$| = t. We say t-set ts is covered by feature set fs iff ts.sel ⊆ fs.sel ∧ ts.$\overline{\text{sel}}$ ⊆ fs.$\overline{\text{sel}}$.

Definition 5 (*Valid t-set*) A t-set ts is valid in a feature model *fm* if there exists a valid feature set fs that covers ts. The set of all valid t-sets for a feature model is denoted with $\mathcal{VTS}^{fm}$.

Definition 6 (*t-wise covering array*) A t-wise covering array tCA for a feature model *fm* is a set of valid feature sets that covers all valid t-sets in $\mathcal{VTS}^{fm}$. Formally, tCA ⊆ $\mathcal{P}$(FS^{fm}) where ∀ts ∈ VTS^{fm}, ∃fs ∈ tCA such that fs covers ts.

Let us illustrate these concepts for *pairwise testing*, meaning when t = 2. From the feature model in Fig. 1, a valid 2-set is [{Driver},{Prim}]. It is valid because the selection of feature Driver and the non-selection of feature Prim do not violate any constraints. As another example, the 2-set [{Kruskal,DFS}, ∅] is

---

[2]An early version is available in [24].
[3]Definitions based on [12, 25].

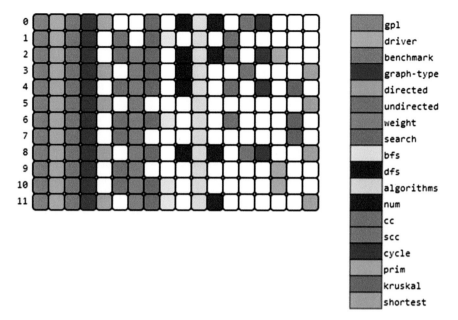

Fig. 2 Graph Product Line 2-wise covering array example [26]

valid because there is at least one feature set, for instance fs1 in Table 1, where both features are selected. The 2-set [∅, {SCC,CC}] is also valid because there are valid feature sets that do not have any of these features selected, for instance feature sets fs0, fs2, and fs3. Notice, however, that the 2-set [∅, {Directed, Undirected}] is not valid. This is because feature GraphType is present in all the feature sets (mandatory child of the root) so either Directed or Undirected must be selected. In total, our running example has 418 valid 2-sets, so a 2-wise covering array must have all these pairs covered by at least one feature set. A covering array can be visually depicted as shown in Fig. 2 [26].

4.2 SPL Genetic Solver (SPLGS)

The *SPL Genetic Solver (SPLGS)* is a constructive genetic algorithm that computes pairwise covering arrays for SPLs based on a feature model that receives as input. It is based on the *Prioritized Genetic Solver* (*PGS*) by Ferrer et al. that takes into account priorities during the generation of test suites [27]. SPLGS extends and adapts PGS for generating test suites of product lines. In each iteration SPLGS adds a new feature set that contributes the most coverage to the partial solution until all

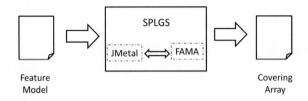

Fig. 3 Architecture of SPLGS

pairwise combinations are covered. SPLGS has been implemented using two
framework tools: (i) jMetal [28], a Java framework aimed at the development,
experimentation, and study of metaheuristics for solving optimization problems;
and (ii) FAMA, an extensible framework for the representation and analysis of
feature models [29]. The architecture of the SPLGS is presented in Fig. 3.

Algorithm 2 sketches the pseudocode of SPLGS. It takes as inputs the feature
model FM. At the beginning, the test suite (TS) is initialized with an empty list
(Line 4), and the set of remaining pairs (RP) is initialized with all valid pairs that
need to be covered (Line 5). In each iteration of the external loop (Lines 6–24), the
algorithm creates a random initial population of individuals (feature sets in our
case) in (Line 8), and enters an inner loop which applies the traditional steps of a
generational evolutionary algorithm (Lines 9–21). That is, some individuals are
selected from the population $P(t)$, recombined, mutated, evaluated, and finally
inserted in offspring population Q. If a generated offspring individual is not a valid
feature set (i.e. it violates any constraint derived from the feature model), it is
transformed into a valid one by applying a Fix operation (Line 15) provided by the
FAMA tool [29]. The fitness value of an offspring individual is the number of pairs
that remains to be covered, and hence it should be minimized (Line 16). In Line 19,
the best individuals of $P(t)$ and Q are kept for the next generation $P(t+1)$. The
internal loop is executed until a maximum number of evaluations is reached. Then,
the best individual found is included in the test suite (Line 22) and RP is updated by
removing the new pairs covered by the selected best solution (Line 23). Then, the
external loop starts again until there is no pair left in the RP set. Finally, in Line 25
the computed test suite is returned. SPLGS has been shown to generate competitive
test suites when compared against other leading CIT approaches for SPL, for further
details refer to [30].[4]

---

[4]In [30] the algorithm is named PGS. We changed its name for this chapter to avoid confusions
with the original algorithm PGS in [27] that was not designed for SPLs.

Algorithm 2. Pseudocode of SPLGS

```
 1: proc SPLGS
 2: Input: FM        // Input feature model
 3: Output: TS       // Output test suite
 4: TS ← ∅   // Empty test suite
 5: RP ← pairs_to_cover(FM)   // Initialize the pairwise configurations
 6: while not empty(RP) do
 7:     t=0
 8:     P(t) ← Create_Population() // P = population
 9:     while evals < totalEvals do
10:         Q ← ∅   // Q = auxiliary population
11:         for i ← 1 to (PGS.popSize / 2) do
12:             parents←Selection(P(t))
13:             offspring←Recombination(PGS.Pc,parents)
14:             offspring←Mutation(PGS.Pm,offspring)
15:             Fix(offspring)
16:             Evaluate_Fitness(offspring)
17:             Insert(offspring,Q)
18:         end for
19:         P(t+1) := Replace (Q,P(t))
20:         t= t + 1
21:     end while   //internal loop
22:     TS ← TS ∪ best_solution(P(t))
23:     RemovePairs(RP, best_solution(P(t)))
24: end while   //external loop
25: return TS
26: end_proc
```

4.3   State of the Art CIT for SPL Testing

There exists an important body of literature on CIT for SPL testing; however, only few examples rely on evolutionary algorithms. In this section we first present these approaches, followed by those that do not rely on evolutionary algorithms. In Sect. 5.4 we present the related work for Multi-Objective Evolutionary Algorithms for SPL testing.

Evolutionary Approaches. Ensan et al. [31] propose a genetic algorithm approach for test case generation for SPLs that uses a variation of cyclomatic complexity metric adapted to feature models and hence their goal is not to provide n-wise coverage. Henard et al. [32] propose an approach based on a (1 + 1) evolutionary algorithm that uses similarity heuristic as a viable alternative for t-wise coverage for coping with large scale feature models and large values of t up to 6. This approach is supported by a tool called PLEDGE that in addition provides a product line editor [33]. Work by Xu et al. [34] uses a genetic algorithm for continuous test augmentation. Their CONTESA tool incrementally generates test cases for branches that have not yet been covered by existing tests. Recent work by Henard et al. [35] creates so called *mutants* of a feature model which in addition to the original feature model are passed to a (1 + 1) evolutionary algorithm to produce test suites.

Non-Evolutionary Approaches. Garvin et al. [36] applied simulated annealing to combinatorial interaction testing for computing n-wise coverage for SPLs. Their algorithm CASA, performs three nested search strategies aiming at iteratively reducing the sizes of the test suites. Perrouin et al. propose an approach that first transforms t-wise coverage problems into Alloy programs and then uses Alloy's automatic instance generation to obtain covering arrays [37]. Oster et al. [38] propose MoSo-PoLiTe, an approach that transforms feature models into Constraint Solver Problems (CSP) to compute pairwise covering arrays. MoSo-PoLiTe can also include pre-selected products as part of the covering arrays. Hervieu et al. [39] follow a similar approach of using constraint programming for computing pairwise coverage. Regarding model based testing, the work by Lochau et al. [40] relates feature models with a reusable test model expressed with state charts to define and analyse feature dependencies and interactions. Cichos et al. [41] proposed an application of the so-called 150 % model, a model with all variable options included, whose goal is to provide complete test coverage for a given coverage criterion. Johansen et al. [25] propose a greedy approach to generate n-wise test suites that adapts Chvátal's algorithm to solve the set cover problem that makes several enhancements, for instance they parallelize the data independent processing steps. Calvagna et al. have developed CITLab [42], a tool for integrating multiple CIT approaches for SPLs.

5   Multi-objective SPL Testing

The approaches presented in Sect. 4 primarily focus on obtaining test suites that achieve complete coverage of the desired t strength. In other words, their single optimization objective is maximizing t-wise coverage. Though useful in many contexts, this single-objective perspective does not reflect the prevailing scenario where software engineers do face trade-offs among multiple and often conflicting objectives that represent technical and economical constraints. In this section, we first present a formalization of SPLs testing as a multi-objective optimization problem and provide a brief motivation example. Then as an example, we describe our exact algorithm to compute the optimal solutions for pairwise testing for coverage and test suite size optimization. And conclude with an overview of related multi-objective SPL testing approaches.

5.1   Multi-objective Optimization Formalization

There exists a wealth of literature in the context of Evolutionary Multi-Objective Optimization [43] and the application of *Search-Based Software Engineering* (*SBSE*) to software testing [44]. In this section we provide the formalization of SPL testing as a multi-objective optimization problem. Our definitions are based on

[45–47] and are generalizations of our previous work for the case of bi-objective pairwise testing [48].

Definition 7 (*Decision space*) The decision space is the set of possible solutions to an optimization problem. In our context, it corresponds to the set of all possible subsets of valid feature sets represented by a feature model fm, denoted as $\mathcal{DS}^{fm} = \mathcal{P}(\mathcal{FS}^{fm})$. A decision vector is an element of the decision space, that is $x \in \mathcal{DS}^{fm}$.

Definition 8 (*Objective functions*) An objective function is a function that represents a goal to optimize, e.g. $f^{fm} : \mathcal{DS}^{fm} \rightarrow \mathbb{N}$.

As examples, let us consider two objective functions:

– Coverage function. We want to maximize the number of t-wise sets covered by a test suite as follows:

$$f_1^{fm} : \mathcal{DS}^{fm} \rightarrow \mathbb{N},$$

$$f_1^{fm}(x) = |covers(x)|,$$

where covers computes the t-wise sets covered by the feature sets of test suite x.
– Test suite size function. We want to minimize the number of feature sets in the test suite. We define this function as follows:

$$f_2^{fm} : \mathcal{DS}^{fm} \rightarrow \mathbb{N},$$

$$f_2^{fm}(x) = |x|.$$

Definition 9 (*Vector function*) A vector function associated to a feature model fm is defined as[5]:

$$F^{fm} : \mathcal{DS}^{fm} \rightarrow \mathcal{OS}^{fm}$$

$$F^{fm}(x) = \left(f_1^{fm}(x), f_2^{fm}(x), \ldots, f_n^{fm}(x) \right)$$

where $\mathcal{OS}$ is the corresponding objective space, in our context is $\mathcal{OS}^{fm} = \mathbb{N}^n$.

Definition 10 (*Objective vector*) An objective vector is the result of applying the vector function to an element of the decision space. Let $x \in \mathcal{DS}^{fm}$, its objective vector u is defined as: $u = F^{fm}(x) = (f_1^{fm}(x), f_2^{fm}(x), \ldots, f_n^{fm}(x))$.

---

[5]For notational brevity we omit on the vector function and the objective vectors the $\mathbb{T}$ that denotes the transpose on vectors.

Pareto dominance is the most commonly accepted notion of superiority in multi-objective optimization because it is the canonical generalization of the single-objective case [45].

Definition 11 (*Pareto dominance*) Let $x, y \in \mathcal{DS}^{fm}$, $u = F^{fm}(x) = (u_1, u_2, \ldots, u_n)$, and $v = F^{fm}(y) = (v_1, v_2, \ldots, v_n)$ for a feature model fm. Let $u \preccurlyeq v$ mean that u is better than v if there is at least one objective i for which $f_i^{fm}(x)$ is better than $f_i^{fm}(y)$, and there are no objectives for which it is worse. Then we say that objective vector u `Pareto-dominates` objective vector v iff $u \preccurlyeq v$ and $v \not\preccurlyeq u$.

Definition 12 (*Multi-Objective SPL n-wise testing problem*) A multi-objective n-wise SPL testing problem for a feature model fm is a 4-tuple $(\mathcal{DS}^{fm}, \mathcal{OS}^{fm}, F^{fm}, \preccurlyeq)$ whose goal is to find a decision vector $x^* \in \mathcal{DS}^{fm}$ such that it minimizes vector function F^{fm}.

Definition 13 (*Pareto optimal decision vector*) A decision vector $x \in \mathcal{DS}^{fm}$ is Pareto optimal iff it does not exist another $y \in \mathcal{DS}^{fm}$ such that Pareto-dominates it, that is $F(y)^{fm} \preccurlyeq F(x)^{fm}$.

Definition 14 (*Pareto optimal set*) The Pareto optimal set P_*^{fm} of a multiobjective n-wise SPL testing problem for feature model fm and its vector function F^{fm} is: $P_*^{fm} = \{x \in DS^{fm} | \nexists x' \in DS^{fm} \text{such that } F^{fm}(x') \preccurlyeq F^{fm}(x)\}$.

Definition 15 (*Pareto front*) For a given multi-objective n-wise SPL testing problem for feature model fm and a Pareto optimal set P_*^{fm}, the Pareto front is defined as: $PF_*^{fm} = F^{fm}(P_*^{fm})$.

5.2 An Example Scenario

Let us now motivate the importance of multi-objective optimization for SPL with a simple and illustrative example. Consider for instance our two objective functions f_1^{fm} and f_2^{fm}, as described above, that respectively represent the t-wise coverage and test suite size. On one hand we want to maximize t-wise coverage while *at the same time* we want to minimize the test suite size. Figure 4 shows the Pareto front for our running example GPL for the case of pairwise testing. Objective function f_1^{fm} is shown on the vertical axis as percentage of coverage pairs, while objective function f_2^{fm} is shown on the horizontal axis.

Taking a multi-objective approach and computing a Pareto front allows software engineers to select not just one solution, as in the case of single-objective techniques, but instead to select from an array the solution that best matches the economical and technological constraints of their testing context. In our example, some

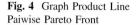

Fig. 4 Graph Product Line Paiwise Pareto Front

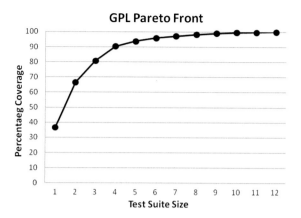

of the questions that can be answered and hence can help software engineers make informed decisions are:

– *What is the minimum size of a test suite that guarantees full pairwise coverage?* Clearly, from Fig. 4, this can only be achieved with 12 feature sets.
– *How many feature sets are needed to get a certain percentage coverage, for example 90 % coverage?* Again, from the information provided by the Pareto front we can affirm that only 4 products are needed to attain 90 % coverage.
– *If only 3 feature sets can be tested because of economical constraints, what is the maximum coverage that can be achieved?* Once more, using the information of the Pareto front, the maximum coverage is 80.86 %.

For this kind of concerns, software engineers not only get a single value, like the number of feature sets to test, but in addition they can also obtain a list of test suites that meet the desired criteria. In sharp contrast with single-objective approaches that can only provide a single solution. For example, the test suite shown in Fig. 2 is just but a single point that is mapped to the GPL Pareto front, in our case the rightmost point in Fig. 4 in the best scenario. Hence, for instance, the questions posed above cannot be addressed with a single-objective method. Next section we present an approach to compute the exact Pareto front which we used for computing the front for GPL.

5.3 Computation of Exact Pareto Fronts

In this section we present an overview of our work on computing exact Pareto fronts for SPL pairwise testing for two objectives, maximizing the pairwise coverage while minimizing the test suite size. For further details please refer to [49]. The algorithm we proposed for obtaining the optimal Pareto set is given in Algorithm 3, and is based on the work of Arito et al. [50] for solving a multi-objective test suite minimization problem in regression testing. From the definitions in the previous subsection, recall that a Pareto optimal set is a set of non-dominated solutions each of which is not dominated by any other solution in the decision space, while the Pareto front is the projection of this set in the objective space, that is, a plot containing the values of the objective functions for each solution.

Algorithm 3 takes as input a feature model (Line 2) and computes the optimal Pareto set (Line 3). It first initializes the optimal set to empty (Line 4). Then it adds to the set two solutions that are always in the set: the empty solution with zero coverage (Line 5) and one arbitrary solution (Line 7) with coverage C_2^f, that is, the number 2-combinations of the set of features (Line 6). The algorithm then enters a loop (Lines 9–15) in which successive zero-one linear programs are generated (Line 12) for an increasing number of products starting at 2. A zero-one program is an integer program in which the variables can take as value either 0 or 1 [51]. In our case, this program serves to compute a solution which has the maximum coverage that can be obtained with i feature sets. In short, we describe this process in more detail.

Each mathematical model is then solved using an extended SAT solver (Line 13), in our case MiniSat+.[6] This solver provides a test suite with the maximum coverage for the given number of feature sets. This solution is subsequently added to the optimal Pareto set (Line 10), and the corresponding coverage is adjusted (Line 14). The algorithm stops when adding a new product to the test suite does not increase the coverage. The obtained Pareto optimal set is finally returned (Line 16).

Now we describe how to build the zero-one program for pairwise coverage. Let n be the fixed number of feature sets we want to compute and let f be the number of features of the input feature model FM. We use the set of decision variables $x_{i,j} \in \{0, 1\}$ where $i \in \{1, 2, \ldots, n\}$ and $j \in \{1, 2, \ldots, f\}$, such that variable $x_{i,j}$ is 1 if feature set i has feature j selected and 0 otherwise. The zero-one program consists of four parts as described next.

1. **Constraints from feature model**. Recall that not all the combinations of features form valid products. The validity of any feature set denoted by a feature model FM can be expressed as a Boolean formula following the standard mapping to Conjunctive Normal Form (CNF) [12]. Each of the CNF clauses is then converted to a constraint of a zero-one program. First, let us define the Boolean vectors v and u as follows [52]:

---

[6]Available at URL: http://minisat.se/MiniSat+.html.

Algorithm 3. Algorithm for obtaining the Pareto optimal set

1: **proc** Pareto Optimal Set
2: **Input:** FM   // Input feature model
3: **Output:** $optimal\_set$ // Output Pareto optimal set
4: $optimal\_set \leftarrow \{\varnothing\}$;
5: $cov[0] \leftarrow 0$;
6: $cov[1] \leftarrow C_2^f$;
7: $sol \leftarrow$ arbitraryValidSolution(fm);
8: $i \leftarrow 1$;
9: **while** $cov[i] \neq cov[i-1]$ **do**
10:    $optimal\_set \leftarrow optimal\_set \cup \{sol\}$;
11:    $i \leftarrow i+1$;
12:    $m \leftarrow$ prepareMathModel(fm,i);
13:    $sol \leftarrow$ solveMathModel(m);
14:    $cov[i] \leftarrow |covers(sol)|$;
15: **end while**
16: **return** $optimal\_set$

$$v_j = \begin{cases} 1 & \text{if feature } j \text{ appears in the clause}, \\ 0 & \text{otherwise}, \end{cases}$$

$$u_j = \begin{cases} 1 & \text{if feature } j \text{ appears negated in the clause}, \\ 0 & \text{otherwise}. \end{cases}$$

With the definitions of u and v, Eq. 1 describes how to write the constraint that corresponds to a CNF clause for the ith product.

$$\sum_{j=1}^{f} v_j(u_j(1-x_{i,j}) + (1-u_j)x_{i,j}) \geqslant 1 \qquad (1)$$

As an illustration, let us suppose in our GPL running example that feature *Search* is the 8th feature in the feature list and *Num* is the 12th feature. The cross-tree constraint "*Num* requires *Search*", shown in Fig. 1, can be written in CNF with the clause ¬*Num* ∨ *Search* and its translation to a zero-one constraint is: $1 - x_{i,12} + x_{i,8} \geqslant 1$.

2. **Constraints for pairwise coverage per feature set**. Because our focus is pairwise coverage, we need to consider four possible combinations between two features: (i) both features unselected, (ii) first feature unselected and second feature selected, (iii) first feature selected and second feature unselected, (iv) both features selected.

We introduce one variable in our program for each feature set, each pair of features and each of these four possibilities. The variables, called $c_{i,j,k,l}$, take value 1 if feature set i covers the pair of features j and k with the combination l.

The combination l is a number between 0 and 3 representing the selection configuration of the features according to the next mapping: $l = 0$, both unselected; $l = 1$, first unselected and second selected; $l = 2$, first selected and second unselected; and $l = 3$ both selected. The values of the variables $c_{i,j,k,l}$ depend on the values of $x_{i,j}$. In order to reflect this dependence in the mathematical program we add the following constraints for all $i \in \{1, \ldots, n\}$ and all $1 \leqslant j < k \leqslant f$:

$$2c_{i,j,k,0} \leqslant (1 - x_{i,j}) + (1 - x_{i,k}) \leqslant 1 + c_{i,j,k,0} \tag{2}$$

$$2c_{i,j,k,1} \leqslant (1 - x_{i,j}) + x_{i,k} \leqslant 1 + c_{i,j,k,1} \tag{3}$$

$$2c_{i,j,k,2} \leqslant x_{i,j} + (1 - x_{i,k}) \leqslant 1 + c_{i,j,k,2} \tag{4}$$

$$2c_{i,j,k,3} \leqslant x_{i,j} + x_{i,k} \leqslant 1 + c_{i,j,k,3} \tag{5}$$

3. **Constraints for pairwise coverage of all feature sets.** Variables $c_{i,j,k,l}$ inform about the coverage in one feature set. We need new variables to count the pairs covered when all the feature sets are considered. These variables are called $d_{j,k,l}$, and take value 1 when the pair of features j and k with combination l is covered by some product and 0 otherwise. This dependence between the $c_{i,j,k,l}$ variables and the $d_{j,k,l}$ variables is represented by the following set of inequalities for all $1 \leqslant j < k \leqslant f$ and $0 \leqslant l \leqslant 3$:

$$d_{j,k,l} \leqslant \sum_{i=1}^{n} c_{i,j,k,l} \leqslant n \cdot d_{j,k,l} \tag{6}$$

4. **Maximization goal.** Finally, the goal of our program is to maximize the pairwise coverage, which is given by the number of variables $d_{j,k,l}$ that are 1. This is expressed
as:

$$\max \sum_{j=1}^{f-1} \sum_{k=j+1}^{f} \sum_{l=0}^{3} d_{j,k,l} \tag{7}$$

In summary, the mathematical program is composed of the goal (7) subject to the $4(n + 1)f(f - 1)$ constraints given by (2) to (6) plus the constraints of the FM expressed with the inequalities (1) for each product. The number of variables of the program is $nf + 2(n + 1)f(f - 1)$. The solution to this zero-one linear program is a test suite with the maximum coverage that can be obtained with n feature sets.

Evaluation. We have evaluated our approach using a benchmark of 118 feature models publicly available in two open repositories [49], whose results are shown in Fig. 5. These feature models have number of feature sets that ranges from 16 to 640. We found that execution time does not grow linearly with the number of feature sets of the

Fig. 5 Time (log scale) to compute Pareto optimal set versus number of feature sets

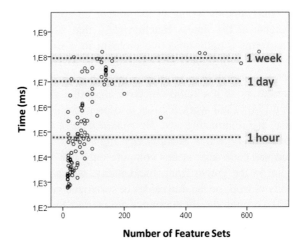

feature models, but instead it grows faster. Consequently we found scalability issues with our approach. Even though the majority of our examples finished within an hour, there were a significant portion that required a day and a few less that required a week of devoted computation in a standard desktop environment. Scalability issues of exact methods are the main reason for using approximate methods based on multi-objective evolutionary algorithms that we summarize in the next section.

5.4 Sate of the Art in Evolutionary Multi-objective Optimization for SPL Testing

We have performed a systematic mapping study on SBSE techniques applied to SPLs [23].[7] In this section we shortly summarize all the works found by this study and in addition describe the salient related work that uses evolutionary multi-objective algorithms for SPLs but not for testing.

Our previous work makes a comparison of four classical multi-objective evolutionary algorithms for SPL pairwise testing, namely: NSGA-II, PAES, MOCell, and SPEA2 [48]. In addition, this work analyzes the performance impact of three different seeding strategies that exploit different levels of domain knowledge to create the initial populations. We evaluated this work using 19 representative feature models from different application domains, ranging in number of features from 9 to 117, and in number of feature sets from 32 to 1,741,824. We found that the algorithms NSGA-II, SPEA2 or MOCell perform comparatively equal and perform best when using the seeding strategy that exploits the most domain knowledge (i.e. seeds the initial population based on a test suite computed using a single-objective algorithm).

---

[7]An early version is available in [24].

The work by Wang et al. present an approach to minimize test suites using weights in the fitness function [53], that is, it uses a *scalarizing function* that transforms a multi-objective problem to a single-objective one [47]. Their work uses three objectives: test minimization percentage, pairwise coverage, and fault detection capability. A similar approach is taken in recent work by Henard et al. that present an ad-hoc multi-objective algorithm whose fitness function is also scalarized [54]. Their work focuses also on maximizing coverage, minimizing test suite size, and minimizing cost. *We should remark that neither of these two approaches are multi-objective evolutionary algorithms in the strict sense.* Clearly, this is because these approaches compute only one single solution, that is, just a single point in the Pareto front. Incidentally, we should point out there is an extensive body of work on the downsides of scalarization in multi-objective optimization (e.g. [55]). Among the shortcomings are the fact that weights may show a preference of one objective over the other and, most importantly, the impossibility of reaching some parts of the Pareto front when dealing with convex fronts.

We should also point out that there is a considerable number of applications of multi-objective evolutionary algorithms but outside of the testing activities of SPL development. A common task where these algorithms is employed is in product configuration. For example, Cruz et al. employ the multi-objective algorithm NSGA-II to create and manage product portfolios based on customer satisfaction and costs [56]. As another example, the work by Sayyad et al. performs a more thorough exhaustive application and analysis of multi-objective evolutionary algorithms for configuration tasks [57, 58]. Our own previous work has also explored using several classical multi-objective evolutionary algorithms for the configuration of dynamic product lines for mobile applications [59].

For sake of completeness, we should also indicate ongoing work on exact multi-objective method by Olaechea et al. who propose an exact method to compute Pareto fronts showing their capability to handle small and medium size problems and provide basic guidelines for choosing either exact or evolutionary approaches [60]. Similarly, Murashkin et al. present a tool for the visualization and exploration of variants in a multi-dimensional space but do not address SPL testing issues [61].

6 Evolutionary Testing of SPLs in Practice

The most common scenario for the development of SPLs in industrial setting comes after the realization that developing and maintaining multiple similar systems, an approach called *"Clone and Own"*, is not economically feasible [62]. The main goals of reverse engineering a SPL from a set of similar software systems are: (i) capture the knowledge of what is common and what is variable (e.g. commonality and variability) in all the artifacts employed throughout the development life-cycle, and (ii) express with a feature model all the valid feature combinations required for the SPL. The result of the reverse engineering process is illustrated in Fig. 6.

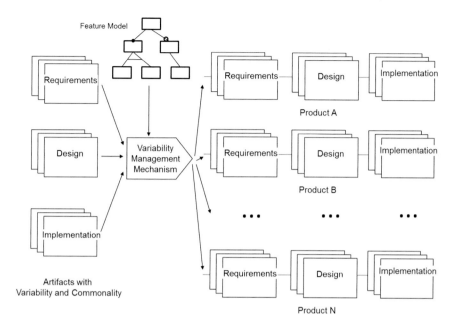

Fig. 6 Overview of reversed engineered SPL

There has been extensive work over the last two decades on how to capture the variability and commonality knowledge for SPLs (i.e. Variability Management Mechanism in Fig. 6), for a summary see for example [63]. Regarding the extraction of feature models, recent work by Lopez-Herrejon et al. also uses evolutionary algorithms in combination of information retrieval metrics for obtaining feature models based on the feature combinations [64, 65]. Alternative approaches can obtain feature models based on generic propositional logic constraints [66].

Once a SPL infrastructure has been put in place, SPL testing can proceed. There has been recent accounts of SPL testing in industrial settings, some of them relying on evolutionary approaches. For instance, Wang et al. report on an experience in the application of multi-objective algorithms for a video conference application [67]. The common trend in such experiences is that the critical factor is eliciting the right feature models from the software engineers, this is so because the information of the valid feature combinations are commonly not well documented if at all. The application of the testing techniques in general do not require expensive hardware or software infrastructure, as in the common cases their execution takes a few minutes or hours in typical off-the-shelf desktop computers. There are, however, empirical and theoretical studies on large scale feature models, mostly of academic interest, that consider large number of features, products, or higher array strengths (e.g. up to 6) which show approaches for coping with scalability issues (e.g. [32]).

7   Open Challenges and Questions

In this section we describe what we consider the most salient open challenges and questions for SPL testing with evolutionary techniques.

Multi-objective optimization. There is an extensive body of research literature in multi-objective optimization that remains largely untapped, for an overview see for example [45, 46]. An open question is whether other multi-objective evolutionary algorithms can yield better results and under which circumstances. In addition, further studies are needed that explore dealing with more optimization objectives which can for instance include information such as control-flow or non-functional properties. Furthermore, it is an open question if so-called *many-objective* optimization algorithms, those that deal with four or more objectives, can also be effectively applied to the context of SPL testing. Of crucial importance is their scalability as the complexity of the feature models or the strength of covering arrays increases.

Need of community-wide testing benchmarks and comparison frameworks. We found that the majority of works employs feature models extracted from common repositories such as SPLOT.[8] However, the selection of which feature models to analyze in each paper appears to be arbitrary most of the times. The first steps towards a benchmark for CIT SPL testing are advocated in [30]. Work by Perrouin et al. proposed a comparison framework which is applied to two different approaches [68]. However, comparisons can only be made per feature model, which makes it infeasible to identify which approach performs overall better or under what characteristics of the feature models [30]. Without a proper and fair benchmark and comparison framework, the progress in the research and its transfer to industry are severely hampered.

Exploiting more SPL domain knowledge. Because of the typically large number of individual systems of a SPL, any information that could be exploited to reduce the search effort is worth of consideration. For example, Haslinger et al. leverage information from feature models to speed up the computation of covering arrays by eliminating redundant *t*-sets [69, 70]. As other examples, the work by Xu et al. that exploits static analysis techniques for achieving coverage more effectively [34], and the work by Lopez-Herrejon et al. that studies seeding strategies [48]. It remains an open question, whether any of the analysis techniques recently surveyed by Thüm et al. (see [71]) could be exploited for this purpose. Also, recent work by Fischer et al. (see [72]) computes traceability links from features and feature interactions to the artifacts that realize them. It is an open question whether such traceability information could also be helpful to prune the search space.

Test suite prioritization. Johansen et al. [73] propose a greedy algorithm that adds weights to products to guide the computation of the t-wise sets. These weights are meant to represent priority values such as commercial importance. An alternative parallel evolutionary algorithm was proposed by Lopez-Herrejon et al. for

---

[8]http://www.splot-research.org/.

this scheme that can produce smaller test suites [74]. Additionally, once test suites have been computed their feature sets can be ordered according to some criteria. For example, Al-Hajjaji et al. propose a prioritization based on similarity [75], while Sánchez et al. compare five SPL-specific prioritization criteria and analyze their effect in detecting faults in order to provide faster feedback and reduce debugging efforts [76]. Another example is the recent work by Wang et al. who proposed a scalarized four-objective function to prioritize quasi-pairwise (not considering all four possible combinations of a 2-wise set) tests suites [67]. Test suite prioritization has a long research literature for single software systems (e.g. [77]) which has not been thoroughly researched within the context of SPLs. Some open research issues are prioritization using combinations of clustering techniques and classical multi-objective evolutionary algorithms (instead of scalarization approaches) that exploit the values obtained for instance from non-functional properties.

Supporting testing and analysis phases of CIT. As mentioned before, CIT consists of four phases [21]. However, the current focus for SPL has mostly been on the first two: modeling relying on feature models, and sampling as summarized in Sects. 4.3 and 5.4. Hence, there is a dire need of research and practice that addresses this limitation. Tools such as EvoSuite[9] could be leveraged as starting point for such tasks.

8   Conclusions

Software Product Lines are an emerging software development paradigm that aims to provide a systematic and methodological reuse of all the assets involved in the development of families of software systems, where products share common functionality but also can have unique distinct features. The proven benefits of SPL practices (e.g. [3]) have resulted in an increasing interest both by researchers and practitioners on effective techniques and tools for adequately testing SPLs. The most important challenge faced is dealing with the generally large number of products, i.e. combinations of features, which makes an infeasible alternative testing individually each one of them.

Recent surveys not only highlight the increasing interest in the area but also several shortcomings and opportunities that exist on the field [4, 5, 7, 16–18]. Within the area of Search-Based Software Engineering a major research focus has been software testing [8, 10, 22], also including evolutionary computation techniques. However, most of the applications are for one-off systems rather than SPLs. The goal of this chapter is to provide an overview of the state of the art in SPL testing and framing evolutionary approaches within that context. We have put forward several challenges and open questions that we believe could be fruitful avenues for further research and practice.

---

[9]http://www.evosuite.org/.

82   R.E. Lopez-Herrejon et al.

Acknowledgments This research is partially funded by the Austrian Science Fund (FWF) projects P 25513-N15, P 25289-N15, and Lise Meitner Fellowship M1421-N15, and by the Spanish Ministry of Economy and Competitiveness and FEDER under contract TIN2011-28194 and fellowship BES-2012-055967. It is also partially founded by projects 8.06/5.47.4142 (collaboration with the VSB-Tech. Univ. of Ostrava) and 8.06/5.47.4356 (Andalusian Agency of Public Works).

References

1. Pohl, K., Bockle, G., van der Linden, F.J.: Software Product Line Engineering: Foundations, Principles and Techniques. Springer, Berlin (2005)
2. Batory, D.S., Sarvela, J.N., Rauschmayer, A.: Scaling step-wise refinement. IEEE Trans. Softw. Eng. **30**(6), 355–371 (2004)
3. van der Linden, F., Schmid, K., Rommes, E.: Software Product Lines in Action—The Best Industrial Practice in Product Line Engineering. Springer, Berlin (2007)
4. Engström, E., Runeson, P.: Software product line testing—A systematic mapping study. Inf. Softw. Technol. **53**(1), 2–13 (2011)
5. da Mota Silveira Neto, P.A., do Carmo Machado, I., McGregor, J.D., de Almeida, E.S., de Lemos Meira, S.R.: A systematic mapping study of software product lines testing. Inf. Softw. Technol. **53**(5), 407–423 (2011)
6. Lee, J., Kang, S., Lee, D.: A survey on software product line testing. 16th International Software Product Line Conference, pp. 31–40 (2012)
7. do Carmo Machado, I., McGregor, J.D., de Almeida, E.S.: Strategies for testing products in software product lines. ACM SIGSOFT Softw. Eng. Notes **37**(6), 1–8 (2012)
8. Harman, M., Mansouri, S.A., Zhang, Y.: Search-based software engineering: trends, techniques and applications. ACM Comput. Surv. **45**(1), 11 (2012)
9. Eiben, A., Smith, J.: Introduction to Evolutionary Computing. Springer, Berlin (2003)
10. McMinn, P.: Search-based software testing: past, present and future. In: ICST Workshops, pp. 153–163. IEEE Computer Society (2011)
11. Kang, K., Cohen, S., Hess, J., Novak, W., Peterson, A.: Feature-oriented domain analysis (FODA) feasibility study. Technical Report CMU/SEI-90-TR-21, Software Engineering Institute, Carnegie Mellon University (1990)
12. Benavides, D., Segura, S., Cortés, A.R.: Automated analysis of feature models 20 years later: a literature review. Inf. Syst. **35**(6), 615–636 (2010)
13. Lopez-Herrejon, R.E., Batory, D.S.: A standard problem for evaluating product-line methodologies. In: Bosch, J. (ed.) GCSE. Volume 2186 of Lecture Notes in Computer Science, pp. 10–24. Springer, Berlin (2001)
14. Michalewicz, Z., Fogel, D.B.: How to Solve It: Modern Heuristics, 2nd edn. Springer, Berlin (2010)
15. Goldberg, D.: Genetic Algorithms in Search, Optimization, and Machine Learning. Addison-Wesley, Reading (1989)
16. da Mota Silveira Neto, P.A., Runeson, P., do Carmo Machado, I., de Almeida, E.S., de Lemos Meira, S.R., Engström, E.: Testing software product lines. IEEE Software **28**(5), 16–20 (2011)
17. Wohlin, C., Runeson, P., da Mota Silveira Neto, P.A., Engström, E., do Carmo Machado, I., de Almeida, E.S.: On the reliability of mapping studies in software engineering. J. Syst. Softw. **86**(10), 2594–2610 (2013)
18. do Carmo Machado, I., McGregor, J.D., Cavalcanti, Y.C., de Almeida, E.S.: On strategies for testing software product lines: a systematic literature review. Inf. Softw. Technol. **56**(10), 1183–1199 (2014)

19. Cohen, M.B., Dwyer, M.B., Shi, J.: Constructing interaction test suites for highly-configurable systems in the presence of constraints: a greedy approach. IEEE Trans. Softw. Eng. **34**(5), 633–650 (2008)
20. Nie, C., Leung, H.: A survey of combinatorial testing. ACM Comput. Surv. **43**(2), 11:1–11:29 (February 2011)
21. Yilmaz, C., Fouché, S., Cohen, M.B., Porter, A.A., Demiröz, G., Koc, U.: Moving forward with combinatorial interaction testing. IEEE Comput. **47**(2), 37–45 (2014)
22. de Freitas, F.G., de Souza, J.T.: Ten years of search based software engineering: a bibliometric analysis. In: Cohen, M.B., Cinnéide, M.Ó. (eds.) SSBSE. Volume 6956 of Lecture Notes in Computer Science, pp. 18–32. Springer, Berlin (2011)
23. Lopez-Herrejon, R.E., Linsbauer, L., Egyed, A.: A systematic mapping study of search-based software engineering for software product lines. Inf. Softw. Technol. J. (to appear)
24. Lopez-Herrejon, R.E., Ferrer, J., Chicano, F., Linsbauer, L., Egyed, A., Alba, E.: A hitchhiker's guide to search-based software engineering for software product lines. CoRR **abs/1406.2823** (2014)
25. Johansen, M.F., Haugen, Ø., Fleurey, F.: An algorithm for generating t-wise covering arrays from large feature models. 16th International Software Product Line Conference, pp. 46–55 (2012)
26. Lopez-Herrejon, R.E., Egyed, A.: Towards interactive visualization support for pairwise testing software product lines. In: Telea, A., Kerren, A., Marcus, A. (eds.) VISSOFT, pp. 1–4. IEEE (2013)
27. Ferrer, J., Kruse, P.M., Chicano, J.F., Alba, E.: Evolutionary algorithm for prioritized pairwise test data generation. InL: Soule, T., Moore, J.H. (eds.) GECCO, pp. 1213–1220. ACM (2012)
28. Durillo, J.J., Nebro, A.J.: jmetal: a java framework for multi-objective optimization. Adv. Eng. Softw. **42**(10), 760–771 (2011)
29. Trinidad, P., Benavides, D., Ruiz-Cortes, A., Segura, S., Jimenez, A.: Fama framework. In: Software Product Line Conference, 2008. SPLC'08. 12th International (Sept.), pp. 359–359
30. Lopez-Herrejon, R.E., Ferrer, J., Chicano, F., Haslinger, E.N., Egyed, A., Alba, E.: Towards a benchmark and a comparison framework for combinatorial interaction testing of software product lines. CoRR **abs/1401.5367** (2014)
31. Ensan, F., Bagheri, E., Gasevic, D.: Evolutionary search-based test generation for software product line feature models. In: Ralyté, J., Franch, X., Brinkkemper, S., Wrycza, S. (eds.) CAiSE. Volume 7328 of Lecture Notes in Computer Science, pp. 613–628. Springer, Berlin (2012)
32. Henard, C., Papadakis, M., Perrouin, G., Klein, J., Heymans, P., Traon, Y.L.: Bypassing the combinatorial explosion: using similarity to generate and prioritize t-wise test configurations for software product lines. IEEE Trans. Softw. Eng. **40**(7), 650–670 (2014)
33. Henard, C., Papadakis, M., Perrouin, G., Klein, J., Traon, Y.L.: Pledge: a product line editor and test generation tool. In: SPLC Workshops, pp. 126–129. ACM (2013)
34. Xu, Z., Cohen, M.B., Motycka, W., Rothermel, G.: Continuous test suite augmentation in software product lines. In: Proceedings SPLC, pp. 52–61 (2013)
35. Henard, C., Papadakis, M., Traon, Y.L.: Mutation-based generation of software product line test configurations. In: SSBSE, pp. 92–106 (2014)
36. Garvin, B.J., Cohen, M.B., Dwyer, M.B.: Evaluating improvements to a meta-heuristic search for constrained interaction testing. Empirical Softw. Eng. **16**(1), 61–102 (2011)
37. Perrouin, G., Sen, S., Klein, J., Baudry, B., Traon, Y.L.: Automated and scalable t-wise test case generation strategies for software product lines. In: ICST, pp. 459–468. IEEE Computer Society (2010)
38. Oster, S., Markert, F., Ritter, P.: Automated incremental pairwise testing of software product lines. In Bosch, J., Lee, J. (eds.) SPLC. Volume 6287 of Lecture Notes in Computer Science, pp. 196–210. Springer, Berlin (2010)

39. Hervieu, A., Baudry, B., Gotlieb, A.: Pacogen: automatic generation of pairwise test configurations from feature models. In: Dohi, T., Cukic, B. (eds.) ISSRE, pp. 120–129. IEEE (2011)
40. Lochau, M., Oster, S., Goltz, U., Schürr, A.: Model-based pairwise testing for feature interaction coverage in software product line engineering. Softw. Qual. J. **20**(3–4), 567–604 (2012)
41. Cichos, H., Oster, S., Lochau, M., Schürr, A.: Model-based coverage-driven test suite generation for software product lines. In: Whittle, J., Clark, T., Kühne, T. (eds.) MoDELS. Volume 6981 of Lecture Notes in Computer Science, pp. 425–439. Springer, Berlin (2011)
42. Calvagna, A., Gargantini, A., Vavassori, P.: Combinatorial testing for feature models using citlab. In: ICST Workshops, pp. 338–347 (2013)
43. Coello, C.C.: Evolutionary multi-objective optimization website. http://delta.cs.cinvestav.mx/ccoello/EMOO/
44. Zhang, Y.: Search Based Software Engineering Repository. http://crestweb.cs.ucl.ac.uk/resources/sbse_repository/
45. Coello, C.C., Lamont, G.B., Veldhuizen, D.A.: Evolutionary Algorithms for Solving Multi-objective Problems, 2nd edn. Genetic and Evolutionary Computation. Springer, Berlin (2007)
46. Deb, K.: Multi-objective Optimization Using Evolutionary Algorithms, 1st edn. Wiley, New York (June 2001)
47. Zitzler, E.: Evolutionary multiobjective optimization. In: Handbook of Natural Computing, pp. 871–904 (2012)
48. Lopez-Herrejon, R.E., Ferrer, J., Chicano, F., Egyed, A., Alba, E.: Comparative analysis of classical multi-objective evolutionary algorithms and seeding strategies for pairwise testing of software product lines. In: Proceedings of the IEEE Congress on Evolutionary Computation, CEC 2014, Beijing, China, 6–11 July 2014, pp. 387–396. IEEE (2014)
49. Lopez-Herrejon, R.E., Chicano, J.F., Ferrer, J., Egyed, A., Alba, E.: Multi-objective optimal test suite computation for software product line pairwise testing. In: ICSM, pp. 404–407. IEEE (2013)
50. Arito, F., Chicano, F., Alba, E.: On the application of sat solvers to the test suite minimization problem. In: Proceedings of the Symposium of Search Based Software Engineering. Volume 7515 of LNCS, pp. 45–59 (2012)
51. Wolsey, L.A.: Integer Programming. Wiley, New York (1998)
52. Sutton, A.M., Whitley, L.D., Howe, A.E.: A polynomial time computation of the exact correlation structure of k-satisfiability landscapes. In: Proceedings of GECCO, pp. 365–372 (2009)
53. Wang, S., Ali, S., Gotlieb, A.: Minimizing test suites in software product lines using weight-based genetic algorithms. In: GECCO, pp. 1493–1500 (2013)
54. Henard, C., Papadakis, M., Perrouin, G., Klein, J., Traon, Y.L.: Multi-objective test generation for software product lines. In: Proceedings of SPLC, pp. 62–71 (2013)
55. Marler, R., Arora, J.: Survey of multi-objective optimization methods for engineering. Struct. Multi. Optim. **26**(6), 369–395 (2004)
56. Cruz, J., Neto, P.S., Britto, R., Rabelo, R., Ayala, W., Soares, T., Mota, M.: Toward a hybrid approach to generate software product line portfolios. In: IEEE Congress on Evolutionary Computation, pp. 2229–2236 (2013)
57. Sayyad, A.S., Menzies, T., Ammar, H.: On the value of user preferences in search-based software engineering: a case study in software product lines. In: Proceedings of ICSE, pp. 492–501 (2013)
58. Sayyad, A.S., Ingram, J., Menzies, T., Ammar, H.: Scalable product line configuration: a straw to break the camel's back. In: ASE, pp. 465–474 (2013)
59. Pascual, G.G., Lopez-Herrejon, R.E., Pinto, M., Fuentes, L., Egyed, A.: Applying multiobjective evolutionary algorithms to dynamic software product lines for reconfiguring mobile applications. J. Syst. Softw. (2015, to appear)

60. Olaechea, R., Rayside, D., Guo, J., Czarnecki, K.: Comparison of exact and approximate multi-objective optimization for software product lines. In: Gnesi, S., Fantechi, A. (eds.) 18th International Software Product Line Conference, SPLC'14, pp. 92–101. Florence, Italy, 15–19 Sept 2014. ACM (2014)

61. Murashkin, A., Antkiewicz, M., Rayside, D., Czarnecki, K.: Visualization and exploration of optimal variants in product line engineering. In: Proceedings of SPLC, pp. 111–115 (2013)

62. Dubinsky, Y., Rubin, J., Berger, T., Duszynski, S., Becker, M., Czarnecki, K.: An exploratory study of cloning in industrial software product lines. In: Cleve, A., Ricca, F., Cerioli, M. (eds.) CSMR, pp. 25–34. IEEE Computer Society (2013)

63. Chen, L., Babar, M.A.: A systematic review of evaluation of variability management approaches in software product lines. Inf. Softw. Technol. 53(4), 344–362 (2011)

64. Lopez-Herrejon, R.E., Linsbauer, L., Galindo, J.A., Parejo, J.A., Benavides, D., Segura, S., Egyed, A.: An assessment of search-based techniques for reverse engineering feature models. J. Syst. Softw. Spec. Issue Search-Based Softw. Eng. (2015)

65. Linsbauer, L., Lopez-Herrejon, R.E., Egyed, A.: Feature model synthesis with genetic programming. In: Goues, C.L., Yoo, S. (eds.) Search-Based Software Engineering—6th International Symposium, SSBSE 2014, Fortaleza, Brazil, 26–29 Aug 2014. Proceedings. Volume 8636 of Lecture Notes in Computer Science, pp. 153–167. Springer, Berlin (2014)

66. She, S., Ryssel, U., Andersen, N., Wasowski, A., Czarnecki, K.: Efficient synthesis of feature models. Inf. Softw. Technol. 56(9), 1122–1143 (2014)

67. Wang, S., Buchmann, D., Ali, S., Gotlieb, A., Pradhan, D., Liaaen, M.: Multi-objective test prioritization in software product line testing: an industrial case study. In: Gnesi, S., Fantechi, A. (eds.) 18th International Software Product Line Conference, SPLC'14, pp. 32–41. Florence, Italy, 15–19 Sept 2014. ACM (2014)

68. Perrouin, G., Oster, S., Sen, S., Klein, J., Baudry, B., Traon, Y.L.: Pairwise testing for software product lines: comparison of two approaches. Softw. Qual. J. 20(3–4), 605–643 (2012)

69. Haslinger, E.N., Lopez-Herrejon, R.E., Egyed, A.: Using feature model knowledge to speed up the generation of covering arrays. In: Gnesi, S., Collet, P., Schmid, K. (eds.) VaMoS, p. 16. ACM (2013)

70. Haslinger, E.N., Lopez-Herrejon, R.E., Egyed, A.: Improving casa runtime performance by exploiting basic feature model analysis. CoRR abs/1311.7313 (2013)

71. Thüm, T., Apel, S., Kästner, C., Schaefer, I., Saake, G.: A classification and survey of analysis strategies for software product lines. ACM Comput. Surv. 47(1), 6 (2014)

72. Fischer, S., Linsbauer, L., Lopez-Herrejon, R.E., Egyed, A.: Enhancing clone-and-own with systematic reuse for developing software variants. 30th International Conference on Software Maintenance and Evolution (2014, to appear)

73. Johansen, M.F., Haugen, Ø., Fleurey, F.: An algorithm for generating t-wise covering arrays from large feature models. In: SPLC (1), pp. 46–55 (2012)

74. Lopez-Herrejon, R.E., Ferrer, J., Chicano, F., Haslinger, E.N., Egyed, A., Alba, E.: A parallel evolutionary algorithm for prioritized pairwise testing of software product lines. In: Arnold, D. V. (ed.) Genetic and Evolutionary Computation Conference, GECCO'14, Vancouver, BC, Canada, 12–16 July 2014, pp. 1255–1262. ACM (2014)

75. Al-Hajjaji, M., Thüm, T., Meinicke, J., Lochau, M., Saake, G.: Similarity-based prioritization in software product-line testing. In: Gnesi, S., Fantechi, A. (eds.) 18th International Software Product Line Conference, SPLC'14, pp. 197–206. Florence, Italy, 15–19 Sept 2014. ACM (2014)

76. Sánchez, A.B., Segura, S., Cortés, A.R.: A comparison of test case prioritization criteria for software product lines. In: ICST, pp. 41–50 (2014)

77. Yoo, S., Harman, M.: Regression testing minimization, selection and prioritization: a survey. Softw. Test., Verif. Reliab. 22(2), 67–120 (2012)

Author Biographies

Dr. Roberto Erick Lopez-Herrejon is currently a senior postdoctoral researcher at the Johannes Kepler University in Linz, Austria. He was an Austrian Science Fund (FWF) Lise Meitner Fellow (2012–2014) at the same institution. From 2008 to 2014 he was an External Lecturer at the Software Engineering Masters Programme of the University of Oxford, England. From 2010 to 2012 he held an FP7 Intra-European Marie Curie Fellowship sponsored by the European Commission. He obtained his Ph.D. from the University of Texas at Austin in 2006, funded in part by a Fulbright Fellowship sponsored by the U.S. State Department. From 2005 to 2008, he was a Career Development Fellow at the Software Engineering Centre of the University of Oxford sponsored by Higher Education Founding Council of England (HEFCE). His expertise is software product lines, variability management, feature oriented software development, model driven software engineering, and consistency checking.

Mr. Javier Ferrer is a Ph.D. candidate. He had his 5-year Combined Bachelor's and Master's Engineering Degree by the University of Malaga. He has also obtained his M.Sc. in computer science (artificial intelligence) and his postgraduate certificate in education by the same university. His research interests are mainly related to metaheuristics optimization techniques. Specifically, he has several publications on the search based software engineering field. Overall, he has more than 20 publications including journal articles, conference papers, and book chapters. Currently his main research line is focused on the evolutionary testing domain.

Francisco Chicano is an associate professor in the Department of Languages and Computing Sciences of the University of Malaga, Spain. He studied Computer Science (2003) and Ph.D. in Computer Science (2007) at University of Malaga, and Physics (2014) in the National Distance Education University. His research interests include the application of randomized search techniques to Software Engineering problems. In particular, he contributed to the domains of software testing, model checking and software project scheduling. He also works on the landscapes theory of combinatorial optimization problems and the application of theoretical results to the design of new search algorithms and operators. He is the author of more than 70 refereed publications, has 3 best paper awards and has served on more than 30 program committees. He has served as Program Chair in the EvoCOP 2015 conference, as Track Chair in GECCO 2013 and GECCO 2015 and as Guest Editor in Special Issues of Evolutionary Computation (MIT), Journal of Systems and Software and Algorithmica. He is frequent reviewer in more than 10 international top journals and during the course 2014/2015 is Faculty Affiliate in the Colorado State University.

Alexander Egyed is a Full Professor at the Johannes Kepler University (JKU), Austria. He received his Doctorate degree from the University of Southern California, USA and previously work at Teknowledge Corporation, USA (2000–2007) and University College London, UK (2007–2008). Dr. Egyed's work has been published at over a hundred refereed scientific books, journals, conferences, and workshops, with over 3500 citations to date. He was recognized as a Top 1 % scholar in software engineering in the Communications of the ACM, Springer Scientometrics, and Microsoft Academic Search. He was also named an IBM Research Faculty Fellow in recognition to his contributions to consistency checking, received a Recognition of Service Award from the ACM, Best Paper Awards from COMPSAC and WICSA, and an Outstanding Achievement Award from the USC. He has given many invited talks including four keynotes, served on scientific panels and countless program committees, and has served as program (co-) chair, steering committee member, and editorial board member. He is a member of the IEEE, IEEE Computer Society, ACM, and ACM SigSoft.

Prof. Enrique Alba had his degree in engineering and Ph.D. in Computer Science in 1992 and 1999, respectively, by the University of Málaga (Spain). He works as a Full Professor in this university with different teaching duties: data communications, distributed programing, software quality, and also evolutionary algorithms, bases for R+D+i and smart cities at graduate and master programs. Prof. Alba leads an international team of researchers in the field of complex optimization/learning with applications in smart cities, bioinformatics, software engineering, telecoms, and others. In addition to the organization of international events (ACM GECCO, IEEE IPDPS-NIDISC, IEEE MSWiM, IEEE DS-RT, ...) Prof. Alba has offered dozens postgraduate courses, multiple seminars in more than 20 international institutions, and has directed several research projects (6 with national funds, 5 in Europe, and numerous bilateral actions). Also, Prof. Alba has directed 7 projects for innovation and transference to the industry (OPTIMI, Tartessos, ACERINOX, ARELANCE, TUO, INDRA, ZED) and presently he also works as invited professor at INRIA, the Univ. of Luxembourg, and Univ. of Ostrava. He is editor in several international journals and book series of Springer-Verlag and Wiley, as well as he often reviews articles for more than 30 impact journals. He has published 80 articles in journals indexed by Thomson ISI, 17 articles in other journals, 40 papers in LNCS, and more than 250 refereed conferences. Besides that, Prof. Alba has published 11 books, 39 book chapters, and has merited 6 awards to his professional activities. Pr. Alba's H index is 41, with more than 8000 cites to his work.

Metaheuristic Optimisation and Mutation-Driven Test Data Generation

Matthew Patrick

Abstract Metaheuristic optimisation techniques can be used in combination with mutation analysis to generate test data that is effective at finding faults and reduces the human effort involved in software testing. This chapter describes and evaluates various different metaheuristic techniques and considers their underlying properties in relation to test data generation. This represents the first attempt to bring together, compare and review ideas and research related to mutation analysis and metaheuristic optimisation. The intention is that by considering these application areas together, we can appreciate and understand important aspects of their strengths and weaknesses. This will allow us to make suggestions with regards to the ways in which they may be used together for maximum effectiveness and efficiency.

Keywords Metaheuristic optimisation · Mutation analysis · Test data generation · Search based software engineering · Fitness function · Hill climbing · Evolutionary optimisation · Swarm intelligence

1 Introduction

Metaheuristic techniques are abstract procedures that can be used to find or generate lower-level heuristics [1]. Although metaheuristic optimisation is not guaranteed to identify the global optimum solution to a problem [2], it is typically able to find a solution that is sufficiently good enough to be practically useful. Metaheuristic optimisation is particularly effective compared to simple deterministic heuristics when the information available is incomplete, imperfect or limited in some way [1]. The key advantage that metaheuristic optimisation has over other techniques is that it is not necessary to specify how to produce an effective test suite in advance [3].

M. Patrick (✉)
Department of Plant Sciences, University of Cambridge, Downing Street,
Cambridge CB2 3EA, UK
e-mail: mtp33@cam.ac.uk

© Springer International Publishing Switzerland 2016 89
W. Pedrycz et al. (eds.), *Computational Intelligence and Quantitative
Software Engineering*, Studies in Computational Intelligence 617,
DOI 10.1007/978-3-319-25964-2_5

Instead, metaheuristic optimisation uses a fitness function to automatically search for optimal solutions from those that are available [4]. It evaluates how good (fit) each potential solution is and takes advantages of patterns in the fitness landscape in order to seek out and identify optimal solutions to complex and challenging problems.

Metaheuristic optimisation can be applied to the task of test data generation by representing the set of test cases being optimised as a search space for the optimisation technique to explore [3]. Most test adequacy criteria can be encoded directly as fitness functions so that every time a testing goal is achieved, the value returned by the fitness function is slightly higher. For example, in order to generate a test suite that exercises all the branches in the program under test, a fitness function can be constructed so that it counts the number of branches [3] that have been covered so far and assesses how close a test input comes to executing each uncovered branch.

This chapter investigates the ways in which different forms of metaheuristic optimisation have been applied to generate test suites that perform well under mutation analysis. Mutation analysis is a stringent and powerful technique for evaluating the ability of a test suite to find faults [5]. It generates a large number of program variants (known as mutants), each of which features a small syntactic change to the logical and arithmetic constructs of the program code. Mutants are designed to be based on faults programmers are likely to make, so that a test suite which detects most of the artificial mutants is expected to perform well against actual faults [6].

It is difficult to represent the results of mutation analysis as a fitness function, as the procedure is more complex than other testing criteria [7]. Not only is it necessary for the execution of some test case to reach point of mutation, but it must also be able to cause a difference at the point of mutation and propagate that difference to the output. A number of different fitness functions are described in this chapter ranging from simple techniques that count the proportion of mutants detected by the test suite [7–9] through to advanced methodologies that encourage individual test cases to reach, effect and propagate a difference to the output [10–12].

A wide variety of metaheuristic optimisation techniques are explored. They range from simple procedures such as the alternating variable method [12, 13] through to nature-inspired algorithms inspired by biological models of population genetics [7, 11] and coordination within self-organising ant colonies [14, 15]. The techniques have been divided into three broad categories: hill climbing techniques search locally for optimal solutions by increasing or decreasing the current values by a certain amount; evolutionary optimisation techniques evolve a population of candidate solutions by selecting, adapting and recombining existing candidates; and finally swarm intelligence techniques, which optimise a diverse range of solutions that can adapt and respond rapidly to changes in the environment as and when necessary.

Testing is an important part of the software development process because, if left undetected, faults can have serious harmful effects [16]. Mutation Analysis may be used to evaluate the fault-finding abilities of test cases and metaheuristic optimisation can apply this information to generate highly effective test suites [5]. There a wide range of ways in which mutation analysis and metaheuristic optimisation can be combined and it is difficult to know which of these techniques to apply. This chapter compares the similarities and differences between the techniques that have been developed so far and attempts to identify some important properties. It is not possible to identify one technique that is always the best, as they are highly dependent on the particular application. However, it is hoped this chapter will serve as a starting point to guide the practitioner in the right direction.

2 Test Data Generation

Software testing is the process of exercising software with a sample of possible inputs, chosen so as to demonstrate its correctness in a convincing manner [25]. Testing is a key part of the development lifecycle because it helps to ensure that programs function as intended. Test data may be generated according to the specification (black-box) or internal structure (white-box) of the software [16]. It is typically too expensive in computation and human effort to apply black-box or white-box techniques to test software exhaustively [16]. Programs have too many paths and potential input values to test them all. As a result, black-box testing is often performed using randomly chosen inputs and white-box testing is considered adequate once a certain set of structural components are covered by the test suite.

Inadequate testing may result in software products that are unsatisfactory or unsafe. For example, during the first Gulf war, a rounding error in a Patriot surface-to-air missile battery led to it failing to identify an incoming Iraqi Scud missile [17]. As a result, 28 American soldiers were killed and many more were injured. In 2012, an undisclosed fault in a high frequency trading program caused a financial services firm (Knight Capital Group) to lose $440 million in 30 min [18]. The firm lost 75 % of the value of its stock in two days and was sold to another trading company four months later. Testing is crucial for detecting failures and mistakes in software.

Developers make various kinds of mistakes, ranging from incorrectly interpreting the specification through to underestimating the usage requirements or just plain typographic mistakes [19]. Developer mistakes are known as faults. More broadly, a fault is defined as an incorrect step, process or data definition within a program [20]. Faults lead to errors in software behaviour. Testing aims to find as many of the faults in a program as possible by executing it with a variety of inputs and conditions so as to reveal errors [16]. Each set of inputs and conditions used in testing is known as a test case and a collection of test cases is called a test suite. Successful test data generation finds faults in the program under test with as few test cases as possible.

Software is difficult to test because it is intangible, unique and highly specialised to a particular purpose [21]. It is estimated that between 30 and 90 % of the labour resources required to produce a working program are spent on software testing [19].

For example, Microsoft employ approximate one test engineer for every developer [22]. Yet, despite this investment many faults are often missed. The Java Compatibility Kit [23] is an extensive test suite developed for the Java Development Kit (JDK), yet there are still thousands of additional JDK bug reports in Sun's bug database [24]. Most programs have too many paths to show that they are all correct [25]. Testing is very expensive and, partly as a result, it is often incomplete.

Test data generation techniques save time and money, as well as improve the standard of testing by creating test suites automatically according to some adequacy criterion [26]. Zhu et al. [27] describe three categories of criteria: Structural testing emphasises the need to exercise particular components in the program code (statements, branches, paths etc.); Error-based testing ensures the input domain is covered thoroughly (e.g. by partition testing); and Fault-based testing (i.e. mutation analysis) aims to detect a range of artificially introduced mistakes in the software.

3 Mutation Analysis

Mutation Analysis is a fault-based testing technique based around the idea that small syntactic changes can be used to simulate actual faults. The concept was first introduced in 1971 by Richard Lipton [28]. Since then, there have been over 400 research papers and at least 36 software tools have been developed [5, 29]. Mutation analysis is considered superior to other testing criteria because it measures a test suite's ability to find faults (of course this depends on how representative the mutants are).

Mutation analysis is supported by the *competent programmer hypothesis* (experienced programmers produce programs that are either correct or very close to being correct) [30] and the *coupling effect hypothesis* (test suites capable of detecting all the simple faults in a program can also detect most of the more complex ones) [31]. Developers may understand how the program should behave and make a small mistake in its implementation, or have a slight misconception about the intended behaviour and carry it through to the implementation. In either case, small syntactic changes are claimed to be sufficient to represent most faults [6].

A *mutant* is a copy of the original program that has had a small syntactic change (known as a *mutation*) made to its logical and arithmetic constructs. Mutations are typically applied one at a time. For example, in Fig. 1, the greater-than inequality of line 1 has been replaced with a greater-than-or-equal-to inequality. A mutant is said to be *killed* by input values that cause it to output a different result to the original program. The mutant in Fig. 1 is killed when the value of 'a' is equal to 10 and the value of 'b' is not equal to zero. Mutation analysis evaluates test suites according to the number of mutants they kill. A test suite is considered to be effective for the program under test if it kills a large proportion of the mutants that are produced.

The proportion of mutants killed by the test suite is known as its mutation score (see Eq. 1). This value may be used to indicate weaknesses, since if the test suite

Fig. 1 A simple syntactic
mutation

```
f (int a; int b) {
1    If (a > 10)
2        {  a:=a+b;  }
3    else
4        {  a:=a-b;  }
5    return a;
}
```

```
f' (int a; int b) {
    If (a >= 10)
        {  a:=a+b;  }
    else
        {  a:=a-b;  }
    return a;
}
```

fails to kill some of the mutants, it is also likely to miss actual faults. Some mutants cannot be killed, since they function equivalently to the original program for every possible input; they are typically removed from the calculation, so that mutation score is correctly scaled between 0 and 1. Mutation scores are a more useful measurement for test data generation than the actual number of faults detected because finding many faults in the program under test may indicate good test data or poor software. In this way, Mutation analysis functions as an independent adequacy criterion that can be used to provide confidence in the quality of software.

$$\text{Mutation score} = \frac{\text{number of mutants killed}}{\text{number of non-equivalent mutants}} \qquad (1)$$

Mutation analysis can be applied iteratively to help improve the quality and efficiency of a test suite (see Fig. 2). The presence of mutants that are not killed by

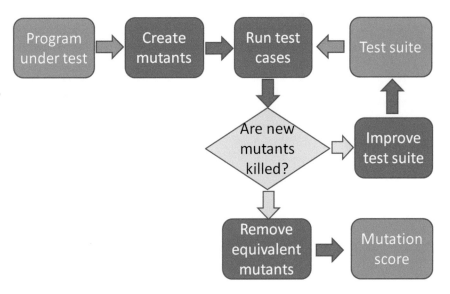

Fig. 2 The process of mutation analysis, adapted from [32]

the existing test suite indicate behaviours of the program under test that are not adequately represented. Test cases may be added or removed in an attempt to kill the remaining mutants as efficiently as possible. Mutation analysis is the reapplied and the test suite is improved until it is able to kill all (or most of) the non-equivalent mutants. The intention is that by generating test data to improve the test suite against a set of mutants, its ability to detect actual faults will also be improved.

4 Metaheuristic Optimisation

Metaheuristic optimisation has been used on a wide variety of problems, from scheduling and planning through to data mining and machine intelligence [2]. It explores candidate solutions to a problem efficiently and identifies optimal or near-optimal solutions. Metaheuristic techniques are not problem-specific [4]. They make few assumptions about the problem being solved, so can be applied 'out of the box' as generalised tools for optimisation. Metaheuristics range from straight-forward search strategies through to advanced methodologies inspired by nature. They can be used to solve complex optimisation problems [4] and have been shown to be effective on problems that involve uncertain, stochastic or dynamic information.

Metaheuristics guide the search procedure through a fitness landscape of candidate solutions (see Fig. 3). For example, each point in the landscape may represent the suitability (or fitness) of a particular test suite. The landscape features peaks and troughs with various heights, gradients and spatial arrangements, as well as other areas where fitness is approximately the same [2]. The aim of metaheuristic optimisation is to guide the search towards the highest possible value in the landscape. In our example, this would correspond to a highly effective test suite.

Fig. 3 A metaheuristic fitness landscape

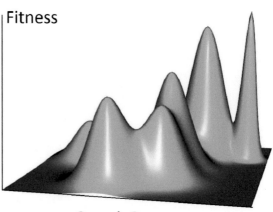

Metaheuristics guide the search using a simple set of rules that describe what to do when particular fitness values are encountered. Typically these rules are stochastic, so that the decisions made and the solutions found are dependent on a set of random variables [4]. Fitness values are not usually calculated in advance, but evaluated as and when needed by the optimisation technique. For example, each test suite can be applied to a set of mutants to generate a mutation score. Metaheuristic optimisation does not guarantee that a globally optimal solution will be found, but it typically finds good solutions with less computational effort than other techniques.

Rather than specifying how to produce an effective test suite, metaheuristic optimisation searches the fitness landscape iteratively, taking advantage of patterns of suitability as it goes along [2]. Figure 4 describes the general processes involved in metaheuristic optimisation. First, one (or a set of) initial candidates is chosen (typically at random). If a candidate meets the fitness termination criteria, the candidate solution is saved and the optimisation process is stopped. Otherwise, further optimisation is performed to create new candidates to evaluate. The process continues until a candidate is found that meets the fitness criteria.

Metaheuristic techniques have to find a balance between intensifying and diversifying their search. Intensification exploits information from previous good solutions to fine tune parameters on a local scale [2]. Diversification explores new regions of the landscape on a global scale and finds solutions different from those found before. Rather than fixing the rates at which diversification and intensification are used, metaheuristic techniques typically alternate between the two approaches dynamically [33]. This allows flexibility to the patterns of fitness they encounter.

A decision must also be made regarding the number of candidate solutions to maintain concurrently. Single solution approaches maintain one candidate at a time

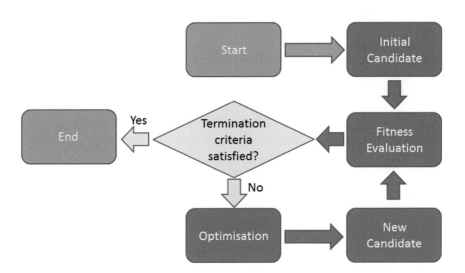

Fig. 4 The process of metaheuristic optimisation

[33], adapting and enhancing its parameter values to form a continuous search trajectory through the fitness landscape. By contrast, population-based approaches maintain multiple candidates [33] and adapt their parameter values simultaneously, so as to evolve a set of points in the fitness landscape. Single solution approaches are typically biased towards intensification [1]. They are effective at fine-tuning parameter values through local search. Population-based approaches typically focus on diversification [1]. They are better suited towards exploring the entire landscape.

5   Using Metaheuristic Optimisation to Kill Mutants

Metaheuristic optimisation is an efficient way to generate and select test suites for mutation analysis. A wide variety of techniques have been used, ranging from hill climbing through to evolutionary optimisation and swarm intelligence. There are also a number of ways to describe and evaluate the fitness landscape. It may be constructed so as to optimise input values for individual test cases or for the entire test suite. This section explores the various metaheuristic techniques and fitness functions that have been used in mutation analysis for test data generation.

There is a danger of introducing new techniques just because they are based upon a new biological metaphor or perform optimisation in a slightly different way, rather than because they are more effective than any other technique. Sörensen [34] argues this could lead metaheuristic research away from scientific rigour. We therefore focus on the general principles, advantages and disadvantages of each technique.

5.1   Fitness Functions Based on Mutation Analysis

The choice of fitness function is important for successful metaheuristic optimisation. Good fitness functions have two key properties: they must be able to differentiate effectively between desirable and undesirable candidate solutions—without this, optimisation may converge to a poor solution, or not converge at all; they must also be inexpensive to calculate—since the fitness function is used extensively during optimisation, if it is too expensive, the optimisation process will become unfeasible.

One option is to evaluate candidate test suites according to their mutation score (see Eq. 1). The higher the mutation score, the fitter the test suite is considered to be. Ghiduk [7] and Mishra et al. [8] apply mutation score as a fitness function in their genetic algorithms for test data generation. Baudry et al. [9] take a similar approach, but with a bacteriological algorithm. Mutation score is a simple and direct fitness function, but it can also be very expensive. Every time a test case is added or changed, it must be run against all of the mutants to find out which ones are killed.

Other fitness functions exist that are cheaper than mutation analysis, but still guide optimisation towards mutant killing test cases. Independent path coverage [35] uses McCabe's Cyclomatic Complexity metric to identify a set of 'basis paths' that can be combined to describe all the paths through a program, then counts how many are covered by the test suite. Mala and Mohan [36] apply independent path coverage as part of their fitness function. It is more stringent than branch coverage and (unlike full path coverage) it is computationally feasible.

$$\text{Independent Path Coverage} = \frac{\text{number of basis paths covered}}{\text{total number of basis paths}} \qquad (2)$$

The techniques described so far are suitable for optimising an entire test suite at once, but sometimes it is more efficient to use other intermediate criteria that guide the optimisation of individual test cases to kill each mutant. With mutation score, it is only possible to know whether a particular mutant is killed. Intermediate criteria may be used to evaluate which of the branches leading up to the mutant are exercised, the size of difference that the mutant introduces in the program state and the depth of its propagation through to the output [3]. Each one of these criteria can be considered individually for a more incremental measure of test suite fitness.

The approach level and branch distance metrics (see Fig. 5) can be used to describe how close a test case is towards causing the program execution to reach a point of mutation [3]. The branch containing the mutated statement is given an approach level of zero, and then it is incremented by one for every branch that could prevent the mutation being reached. This means branches with the highest approach level are farthest from the point of mutation. Branch distance measures the difference between the evaluated and required value at the first branch condition where

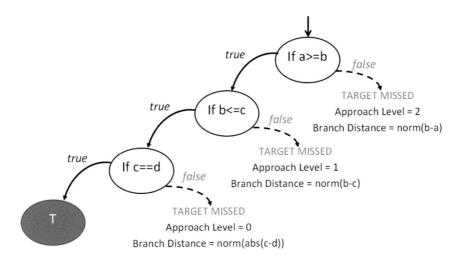

Fig. 5 Example of approach level and branch distance

execution diverts from the intended path [3]. Together, approach level and branch distance can be used to guide test cases towards reaching each point of mutation.

Another metric is needed to describe how far the test data is away from causing a difference at the point of mutation. Papadakis and Malevris [12] measure this using a metric called mutation distance, which is based on the previously described metric for branch distance. For example, consider the original expression $a > b$. If it is mutated to $a \geq b$, the mutation distance is $abs(a - b)$, if it is mutated to $a < b$ it is infinite and if it is mutated to $a \leq b$ it is zero. When combined, the approach level, branch distance and mutation distance satisfy a condition known as weak mutation.

For strong mutation to be achieved, the effects of each mutant must be propagated to the output. Bottaci [10] measures propagation in terms of the number of unequal state pairs that occur between a mutant and the original program once the mutation point is reached. It can be difficult to synchronise the sequence of state pairs when a mutant causes the path to diverge. This is because path divergence does not guarantee state divergence and the program may return back to the original path and state at some later point. It is also prohibitively expensive to keep track of and compare all the program states that result from test data generation. This is why Ayari et al. [15] only implement Bottaci's metrics for reaching and causing a difference at the point of mutation in their optimisation technique for test data generation.

Fraser and Zeller [11] measure the potential for a test case to propagate the effect of a mutant in terms of its impact on statement coverage and function return values. This alleviates the previous problems regarding synchronisation because it does not matter in which order the statements are executed. As well as considering how the impact of a mutant is affected by the choice of test case, we can also consider how the execution of each node in the program affects the mutant's impact. Certain nodes are more likely to propagate the effect of a mutant to the output than others. For every node following the point of mutation, Papadakis et al. [12] count the number of mutants that are killed when that node is executed. The more mutants that are killed, the greater the impact that particular node is considered to have.

Clearly there is a trade-off between the expressive capability and cost of the criteria used to evaluate fitness. Simple criteria such as branch coverage are computationally inexpensive, but produce test suites that are inefficient at killing mutants. Smaller test suites may be produced that achieve a higher mutation score by incorporating propagation into the fitness criteria, but this is typically more expensive. Kaur and Goyal [14] improve the efficiency of their generated test suites by maximising the ratio of mutants killed to execution time (see Eq. 3). However, it is also important to consider the human effort involved in evaluating and understanding the outputs produced by a test suite. One way to do this is to specialise the tests so that they kill specific mutants. Patrick et al. [45] apply the fitness function in Eq. 4 to target each test case at a different group of mutants. Ultimately, human effort is more costly than computational expense and test data generation techniques are more attractive if they reduce the time required to evaluate the test cases.

$$\text{Test case quality} = \frac{\text{number of mutants killed}}{\text{test case execution time}} \tag{3}$$

$$\text{Test case specialisation} = \sum_{m \in M} \sum_{s \in S} \frac{(\bar{K}_m - \bar{K})^2}{(K_{s,m} - \bar{K}_m)^2}$$

$$\bar{K}_m = \left(\sum_{s \in S} K_{s,m} \right) / |S| \qquad \bar{K} = \left(\sum_{m \in M} K_{s,m} \right) / |M| \tag{4}$$

(S is the set of test suites, M the set of mutants, $K_{s,m}$ is the number of times s kills m, $\bar{K}_m$ is the average number of times m is killed and $\bar{K}$ is the average number of times any mutant is killed).

5.2 Hill Climbing

Hill climbing is a simple, yet powerful technique for metaheuristic optimisation. It is considered to be a form of local search [37] because it only explores candidates in the neighbourhood of the current solution. Hill climbing is not guaranteed to find the best possible solution across the entire landscape, but it does quickly find a solution that is locally optimum [37]. In many cases, this is sufficient. Other, more advanced techniques may give better results under certain circumstances, but hill climbing often performs equally well [38] and the algorithm is easier to understand [37]. As such, hill climbing is typically used as a starting point for optimisation research and a benchmark with which to compare other more complicated algorithms [38]. Hill climbing is simple to implement, easy to understand and yet surprisingly effective.

Hill climbing starts with an initial candidate solution (typically selected at random), then updates its parameter values through a series of iterations. At each step, one of the parameter values is changed to a new value in the neighbourhood of the current solution. Hill climbing takes advantage of local gradients in fitness by selecting neighbouring candidates if they improve the fitness evaluation. The candidate's values are repeatedly adjusted until no further improvement in fitness is obtained.

A decision must be made as to the neighbourhood of candidate solutions to evaluate. Once the neighbours are evaluated, hill climbing moves to the neighbour with the highest fitness and a new neighbourhood is created. The simplest strategy is to adjust the value of each parameter systematically (e.g. from left to right), then select the first such move that improves the fitness evaluation [38]. By contrast, steepest ascent hill climbing evaluates the fitness of all possible moves in the neighbourhood, then selects the one that provides the greatest improvement [38]. The steepest ascent strategy takes more time to find the best possible move at each step, but it typically requires fewer steps to reach an optimal solution than simple hill climbing.

The solution found by both the above strategies is only guaranteed to be locally optimal. To increase the chance of finding the global optimum, it may be better not to evaluate every possible move. Stochastic hill climbing selects neighbours to evaluate at each step sparsely according to a probability distribution. In an experiment with combinatorial optimisation, stochastic hill climbing found an optimum faster than both simple and steepest-ascent hill climbing [39]. It was also shown to perform better than two Genetic Algorithms, with populations of 128 and 1024 respectively [39]. This confirms the findings of Harman and McMinn [40] that even though hill climbing is straightforward to implement and easy to understand, it is still effective at test data generation compared to other more complex optimisation techniques.

Hill climbing is often applied to test data generation, using the Alternating Variable Method (AVM), a technique first introduced by Korel [13]. AVM simplifies the process of optimising a test case by adjusting each input parameter in isolation from the rest. Hill climbing is applied to the input parameters one at a time in turn. Whilst hill climbing is being applied to one of the input parameters, the values for all other parameters remain fixed. If adjusting the value for the first variable does not improve the fitness evaluation, the algorithm tries adjusting the next variable, and so on, until every parameter value in the test case has been adjusted.

Adjustments to parameter values are typically made using a combination of exploratory and pattern moves. Exploratory moves change the value of a parameter by a small amount. They are used to determine a suitable direction for the hill climb. If one of the exploratory moves increases the fitness evaluation, then the value that gave the improvement will be used as the new candidate solution. A pattern search is made in the direction of improvement, applying increasingly larger changes to the chosen variable as long as a better solution is found. If none of the neighbours have a higher fitness than the current solution, AVM continues to perform exploratory searches on the other parameters, until either a better neighbour has been found or all the parameters have been unsuccessfully explored.

Papadakis and Malevris [12] applied AVM to kill mutants from 8 small Java programs. Test inputs were optimised, according to a number of different control flow criteria; first so that they reach each point of mutation, then so that they kill the mutants located at each point. The results showed AVM to be more effective than random testing, but only by a small amount. There was little or no difference in effectiveness with fewer than 4500 fitness evaluations [12]. With 50,000 evaluations, AVM killed on average 22 % more mutants than random testing. However, AVM was significantly more effective for two programs (variants of the Triangle program). Arcuri [41] performed a full run-time analysis of AVM on the triangle program. The program performs a simple calculation to determine whether a triangle is a valid equilateral, isosceles or scalene. Hill climbing and the Alternating Variable Method may perform less well may on non-numerical and more complex programs.

In particular, hill climbing performs poorly on fitness landscapes where there are a number of local (i.e. non-global) optima (see Fig. 6). Once optimisation reaches a

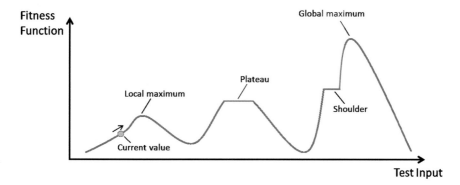

Fig. 6 A challenging landscape for hill climbing

local maximum fitness value, all the neighbouring solutions have lower fitness, so an exploratory search is likely to fail [37]. Similarly at a plateau, the neighbours all have the same fitness and there is no indication as to which direction for the hill climb to travel. The result is that hill climbing often becomes stuck at a local optimum, terminates prematurely, or (in the case of a plateaux) is left wandering aimlessly through the landscape. Success depends largely on the initial conditions, which are determined by the starting point for the hill climb [38]. For example, the global maximum in Fig. 6 has a plateau on one side, which forms a shoulder. This makes the global optimum much harder to reach from one direction than the other. One way to address this problem is to restart the hill climb from a number of different locations (usually chosen at random) to achieve a new maximum fitness value each time. The fitness values are compared and the highest one selected, so as to identify a local optimum that is closer to the global one.

Another option (where available) is to initialise hill climbing using pre-existing test cases. Most test data generation techniques start from scratch, assuming that no test cases already exist [42]. Yet, this is seldom the case. Throughout the various stages of a project, people gather test data for a a variety of different purposes: Requirements engineers capture and explain the ways in which the software is intended to behave with use-cases; software designers expand upon this information using interaction diagrams to describe the flow of information through the system; and programmers use simple test cases to check their code is functioning as they work on it. In addition, test case may be available from testing that was performed on a previous version of the program (these are known as regression tests). Even if this data has been lost, or was not created, it is still possible to generated initial tests used some automated tool, such as Dynamic Symbolic Execution [42]. Overall, there is typically no need to start testing a program from scratch.

Yoo and Harman [42] applied this idea to generate test cases for methods from two real world Java libraries: *binarySearch* is an implementation of the binary search algorithm from a scientific computing library developed at CERN; whereas *read_number* is a numerical parser taken from an Internet event notification library.

They started with a small initial test suite for each program, manually constructed so that it exercises each branch in the code, and allowed their testing tool to search for new test cases by making simple changes to the existing tests. Yoo and Harman's technique achieved mutation scores that were 3.1 and 22.2 % higher for *binarySearch* and *read_number* respectively, when compared to another test data generation tool (Iguana) that uses the more popular alternating variable method.

Pre-existing test cases can be exploited to reduce the time and effort involved in test data generation and produce more effective tests [42]. These test cases are a suitable starting point for hill climbing, since they are likely to be closer to a global optimum than test cases chosen at random. Although testing cannot be used to prove that a program is correct, some confidence may be provided in the success of a large number of test cases. Techniques, such as hill climbing, that apply a local search to existing test data, can be used inexpensively to generate further test cases for evaluation. This makes it less likely for the test cases to be over-specialised to the mutants they were created for, which is useful in ensuring repairs to a program do not allow test cases to pass without fixing the underlying problems [42].

In summary, hill climbing has been shown to be an effective technique for generating test data to kill mutants. It is particularly efficient at fine tuning test cases, but depending on the starting conditions, it is prone to becoming stuck in a local optimum. For this reason, hill climbing is suitable for being used with multiple restarts or combined with some other, more global, optimisation techniques.

5.3 Evolutionary Optimisation

Evolutionary optimisation techniques are inspired by the process of evolution in nature [43]. Candidate solutions compete for the rights to survive and reproduce. Yet, only the fittest candidates are allowed to proceed to the next generation. This leads, over a number of generations, to individuals that are highly adapted to their environment. Evolutionary optimisation techniques are able to find the global optimum even when the fitness landscape is large and complex [44]. For example, they perform well when the fitness function is discontinuous, noisy, changes over time, or has many local optima [43]. This makes evolutionary optimisation techniques suitable for the task of generating test data for large, complex programs that have many different input parameters or types and an intricate system of control and data flow.

As with stochastic hill climbing, evolutionary optimisation applies a randomised process of trial and error to seek for optimal solutions [43]. However, in contrast to hill climbing, evolutionary optimisation is non-local and it maintains a population of multiple candidates at the same time. Non-local optimisation searches a wider range of values in the fitness landscape and is less likely to get stuck in a local optimum. Population-based techniques are intrinsically parallel in nature [44]. It is not necessary to restart the optimisation process when a member of the population fails to reach an optimum, as useful information is preserved in the other candidates.

Since the population of an evolutionary optimisation technique is fixed in size and candidates are selected to be replaced at random, some highly fit individuals are inevitably lost from the population [43]. Although it seems this could cause problems, it actually improves the optimisation. Losing highly fit (and potentially over-trained) candidates allows the search to descend into valleys in order to reach different (and potentially higher) optima [43]. This makes evolutionary optimisation more likely to find values that are closer to the global optimum than hill climbing.

Algorithm 1 presents an overview of the algorithm used by evolutionary optimisation. The initial population is typically chosen at random. Then as long as the termination condition has not been reached, the candidate solutions are evaluated and the fittest individuals are selected to continue into the next generation, after first being adapted and/or recombined according to the particular evolutionary technique.

Algorithm 1 Algorithm for Evolutionary Optimisation

 1: Generate the initial population (typically at random)
 2: **repeat**
 3:     Evaluate fitness values for each candidate solution
 4:     Candidate solutions compete to continue into the next generation
 5:     Select the fittest candidates (allowing for some randomisation)
 6:     *Allow the fittest selected candidate solutions to remain unchanged**
 7:     Adapt the selected candidates by changing some of their values at random
 8:     *Combine values from the selected candidates to create new candidates**
 9:     Form a new population from the various modifications of candidates in the previous one
10: **until** Some termination condition is reached (e.g. fitness above certain level)

---

* Steps 6 and 8 are not present in all evolutionary optimisation algorithms,
they are used by elitist and recombinatorial techniques respectively

Ghiduk [7] represents a candidate solution of test cases using a binary vector of ones and zeroes. The length of this vector is set according to the number and precision of input values to the program under test. Initially the value of each bit is set at random, but they are adjusted throughout the optimisation process and selected according to how much they improve the mutation score of the resulting test suite [7]. Optimisation is terminated when the maximum number of generations (100) is reached or there are no further improvements in mutation score. In experiments with 13 Java programs, evolutionary optimisation achieved a mutation score of 81.8 %, compared to the 68.1 % achieved by random testing.

The encoding of candidate solutions is important as it has a significant impact on the optimisation process. Binary vector representations provide a large number of ways in which candidate solutions can be modified, but some changes are less productive than others and this can make it take longer to reach an optimum. For example, changing the most significant bit of a numerical value has a very different effect to changing the least significant bit. Fraser and Zeller [11] introduce an alternative test case representation using a sequence of method calls. The sequence may be modified by the optimisation technique through the deletion, insertion or modification of program statements. In addition, Fraser and Zeller [11] use a more sophisticated fitness function, using control flow criteria in a similar way to Papadakis and Malevris [12]. The new technique was evaluated against manually devised test suites for two large Java libraries: Joda-Time and Commons-Math. Evolutionary optimisation produced fewer tests and killed more of the mutants from Joda-Time, but produced more tests and killed fewer mutants from Commons-Math.

There are two main driving forces in evolutionary optimisation: variation and selection [43]. At each generation, new candidate values are created by making random adaptations to the parameter values of the existing candidates. This ensures that the diversity of the population is maintained and allows new parameter values and combinations of values to be explored. Also at each generation, allowing for some randomisation in the selection process, the fittest candidates are generally selected and the weaker candidates are removed. This means that as optimisation progresses, there should be a trend towards increasingly fitter candidate solutions. The strengths of the existing candidates in the population are exploited, as their parameter values are carried through into the next generation.

The challenge in setting up an effective evolutionary optimisation technique is to balance the processes of exploration and exploitation so as to achieve an efficient trade-off between them [44]. If the technique concentrates on making the best use of the candidates that are currently available, it might not be able to reach the global optimum. Yet, if it spends time searching for other (potentially more effective) candidates, there is no guarantee that a fitter candidate will be found. Either way, the technique risks wasting time and resources that may be better applied in a different way. Mishra et al. [8] propose the use of an elitist GA to evolve test cases for each unit, whilst maintaining the test cases that have been shown to be particularly effective, in case they are effective on other units. However, it can be difficult to determine in advance for any given situation, which strategy will be the most effective, as this depends on both the program under test and the type of evolutionary algorithm that is used. In practice, it is useful to implement a small trial to compare a number of different options when starting a new project.

A further decision must be made as to whether to include recombination as part of the evolutionary optimisation technique. Recombination is used to combine parameter values from different candidates in the hope that the new combination of values will have the strengths of the previous candidates without any of their weaknesses [44]. In contrast to adaptation, which takes a single candidate solution and changes its parameter values to produce a new candidate, recombination takes two or more candidate solutions and selects parameter values from one or more of the previous candidates [43]. This allows a transfer of information in that candidates can benefit from what other candidates have learnt [44]. However, there is no guarantee that combining parameter values from different solutions will be effective—the values may only be effective in the context of the other parameter values in the original candidate. Recombination can be destructive as well as beneficial, so it is important to consider its role in an evolutionary optimisation technique carefully.

Without recombination, candidates explore the fitness landscape on their own, without interference from the other candidates, but with recombination they are able to take of things other candidates have discovered [44].

Harman and McMinn [40] claim that hill climbing outperforms evolutionary optimisation in its ability to generate effective test suites, as recombination is often disruptive to the optimisation process rather than helpful. Forrest et al. [39] claim that certain 'royal road' functions are particularly suitable for recombination because they allow different sections of the solution to be evolved individually then combined together to form the global optimum. However, it should be noted that since these examples are artificially constructed to be suitable for recombination, it is unclear how frequently these kinds of functions occur in practice. Harman and McMinn [40] claim that recombination should be avoided in most circumstances, so they prefer the use of the Alternating Variable Method for hill climbing.

Nevertheless, recombination does not necessary need to be included as part of an evolutionary optimisation technique and there are a number of ways in which local search can be combined with evolution so as to take advantage of the benefits of each approach. For example, Mala and Mohan [36] use a memetic algorithm to generate test suites. Memetic algorithms apply a local search at each generation to improve the fitness by exploring the immediate neighbourhood of the current candidate. This allows evolutionary optimisation to take advantage of local gradients in fitness as well as searching on a global scale. Mala and Mohan [36] evaluated 18 Java programs from industry and academia and found that they were able to achieve a similar mutation score to a genetic algorithm, but with fewer test cases. Other options include the CMA-ES algorithm, which uses a multivariate Gaussian distribution to describe the neighbourhood of the current best solution in a way that is a compromise between local and global search. Patrick et al. [45] applied a CMA-ES along with dynamic transformation of the program under test to identify effective subdomains of test input. In a study with 6 Java programs, they achieved a 160 % improvement in mutation score compared to random testing.

5.4 Swarm Intelligence

Swarm intelligence techniques are also inspired by a biological concept [46]. However, instead of being based on the inheritance of genetic information, they focus on the ways in which individuals cooperate by sharing information. For example, ants decide where to forage using networks of pheromone trails [47]. If ants encounter an obstacle (see Fig. 7), they look for ways around it at random. However, when some of the ants find a way around it, the other ants follow their pheromone trail to form a new route. There is a genetic element to the coordination of populations (e.g. polymorphism in ants), but the most significant factor in cooperation is self-organisation [46]. Self-organisation refers to the spontaneous way in which coordination arises at the global scale out of local interactions between organisms that are initially disorganised [49]. Individual organisms exhibit

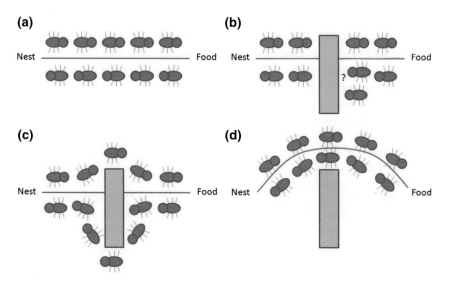

Fig. 7 Ants following pheromones to find their way around an obstacle

simple behaviour and when viewed in isolation, their actions appear noisy and random. However, when multiple organisms work together, complex collective behaviour emerges.

Self-organisation uses simple actions as seeds for random growth in order to create complex behaviour among the collective [48]. It is the way in which the simple actions of individuals interact that allows the collective to search for effective new solutions to a problem. Two forms of interaction exist: positive feedback and negative feedback [49]. Positive feedback reinforces actions that lead to a useful result. For example, when a foraging bee brings nectar back to the hive, it will choose from one of three options: If the bee has found a good source of nectar, it will dance to indicate its direction to other bees; if the source of nectar is mediocre, the bee will return back to the source but not dance; or, if the bee has found a poor source of nectar, it will abandon the source and follow the other bees [48].

Negative feedback acts as a counterbalance to positive feedback by dissuading individuals against making less useful actions [49]. For example, when a source of nectar is exhausted, or merely saturated from too many bees seeking after it at once, the bees will stop dancing so as to prevent other bees from travelling to it [48]. Even when a suitable source of nectar has been found, competition with other sources may prevent bees from travelling to it. Typically, the vigour with which a bee dances is in proportion to the quality of the source it has encountered, so that the majority of bees travel to the best source of nectar [48]. In this way, negative feedback helps to stabilise the behaviour of the collective. Together with positive feedback, negative feedback allows the colony to maintain an optimal supply of nectar to the hive.

Kaur and Goyal [14] use mutation analysis and an artificial bee colony to select and prioritise test cases for regression testing. They evaluate their approach on two C++ programs, for which test suites have been manually developed: a college system for managing course admissions (which has 35 test cases); and a Hotel Reservation system (which has 40 test cases). Kaur and Goyal [14] aim to select a subset of these test cases that form an optimised test suite. Two types of 'bees' are employed: scout bees apply a global search to explore potential candidate test suites and evaluate their fitness according to mutation analysis; by contrast, forager bees start at the fittest test suites that were observed by the scout bees and apply a local search to exploit the neighbourhood of each candidate. Test cases are selected such that they detect faults not detected by the test cases already selected. As a result, test suites can be ordered such that they kill more mutants in less time [14].

Ayari et al. [15] use ant colony optimisation to generate test suites that achieve a high mutation score. The technique starts with a global search performed by 'ants' who evaluate the fitness of random test cases according to how far away they are from killing a mutant. This distance is measured in terms of the number of critical decision nodes that are not traversed and the difference between the current and required value at the node where execution deviates from the path to the mutant [15]. Subsequent ants follow the pheromone trails left by previous ants and perform a local search to take advantage of previous fitness evaluations. Pheromone trails guide ants in constructing test cases, one parameter value at a time. At each step, the ant selects a value that was previously evaluated, or chooses a new value, in proportion to the fitness of the corresponding test cases. Ayari et al. [15] evaluate their approach using two programs written in Java: the Triangle program (as described before); and NextDate, which validates the date that is input and determines the date of the next day. Ant colony optimisation is able to kill more than twice as many mutants as a genetic algorithm and more than three times as many as hill climbing.

Artificial bee colony and ant colony optimisation require the fitness landscape to be described in a discrete way. This makes it difficult to use these algorithms for continuous optimisation tasks, such as generating test cases for programs with numerical inputs. Kaur and Goyal [14] try to get around this problem by generating test cases in advance and optimising the order in which they are executed. Whereas, Ayari et al. [15] discretise the fitness landscape into a limited number of values for each input parameter (spread evenly over the input domain) and then optimise the combination of those input values. These options both help to simplify the optimisation process, and it may be worthwhile to choose them for that reason alone. Yet other models of swarm intelligence (based on particle swarm optimisation) can be applied directly to continuous optimisation problems. Two of these techniques are reviewed below: artificial immune systems and bacteriologic algorithms.

May et al. [50] implement an artificial immune system to generate test suites that are efficient at killing mutants. Artificial immune systems optimise antibodies that are effective against specific antigens. In this case, each antibody represents a test case and each antigen represents a mutant [50]. Test cases are optimised so that they kill at least one mutant not killed by any of the test cases stored in memory as antibodies. The collection of antibodies in memory at the end of the optimisation process are returned to the user as a test suite. The mechanism used to search for new test cases that are effective against the remaining mutants is known as clonal selection theory [50]. Clonal selection theory describes the way in which antigens activate specific antibodies according to their affinity. These antibodies then multiply in numbers by cloning and adapt to be even more effective against the antigen by a process of mutation and selection. May et al. [50] consider test cases that kill more mutants to have a higher affinity (or fitness). New test cases are mutated from the previous ones according to their affinity. High affinity antibodies generate more clones than low affinity ones, but mutate less (as they are closer to the desired solution). May et al. [50] show the artificial immune system is able to kill more mutants with fewer fitness evaluations than a genetic algorithm on four Java programs.

Baudry et al. [9] use a bacterial foraging algorithm to create an effective test suite for a C# parser. Bacteria locate food sources in their environment by sensing and following chemical gradients. They propel themselves along the gradients using long thin structures called flagella. Baudry et al. [9] interpret improvements in mutation score as gradients in food sources and model individual test cases as bacteria that are travelling towards them. Each movement of a bacterium is implemented with a small change to one of the input parameters. Test cases are selected according to their mutation score and the best test cases are allowed to remain within the new population [9]. By remembering which candidates achieve a high mutation score, it is not necessary to recalculate the mutation score for every individual in each generation. Baudry et al. [9] showed that their bacterial foraging algorithm achieved a higher mutation score than a genetic algorithm with fewer mutant executions.

Baudry et al.'s implementation differs from the classic version of a bacterial foraging algorithm [51], in that bacteria only have one mode of movement. In the original paper, Passino [51] described the way in which bacteria move by either swimming or tumbling. Initially, bacteria have no way of knowing which way to travel in order to reach a food source, so they move chaotically through their environment (tumbling). However, once a bacterium is able to detect a gradient, it starts to travel quickly towards the food source (swimming). This can be implemented by dynamically modifying the step size by which parameter values are adapted according to the strength and direction of the fitness gradient. This approach is similar to the multivariate Gaussian adaption used by CMA-ES and the combination of pattern and exploratory search moves in the alternating variable method.

5.5 Comparing Metaheuristic Techniques for Killing Mutants

Table 1 summarises the empirical studies that have been discussed in this chapter. Although it contains a significant proportion of the literature, the table is not intended to be definitive. There are other techniques that have been used to generate test data for mutation analysis which are not included. These techniques, such as adaptive random testing [54] and dynamic symbolic execution [52, 53], are outside the scope of this chapter. The studies that have been included were chosen to demonstrate are broad sample of techniques for metaheuristic optimisation.

Table 1 contains two studies on hill climbing (HC), four studies on evolutionary optimisation (EO) and four studies on swarm intelligence (SI). The studies were primarily conducted in Java, although C, C++, C# and Eiffel have also been used. Table 1 also lists the number of programs included in each study, as well as the total code length (in classes or in lines of code) and the number of mutants generated from the programs under test. Question marks indicate details that were not provided. The studies are arranged in order of the number of mutants (ranging from some unknown number up to 117,913). The latter study included 1416 classes and, although not directly specified, it probably includes the largest number of lines of code as well.

Empirical studies typically compare the performance of the technique they propose against a benchmark of random testing. Papadakis and Malevris [12] showed with 50,000 test cases that hill climbing killed on average 22 % more mutants than random testing [12]. Similarly, Ghiduk [7] showed with 1000 test cases that evolutionary optimisation achieved a mutation score of 81.8 %, compared to the 68.1 % achieved by random testing. Finally, Kaur and Goyal [14] showed that swarm intelligence selected 35 and 40 test cases respectively for two different programs in an order that killed more mutants in less time than random testing. We can therefore state with reasonable authority that metaheuristic optimisation techniques can be more efficient than random testing. Metaheuristic optimisation can be

Table 1 Summary of empirical studies

Study	Technique	Language	#Programs	Code length	#Mutants
Kaur and Goyal [14]	SI	C++	2	NA	NA
Mala and Mohan [36]	EO	Java/C++	18	305 (classes)	NA
Ayari et al. [15]	SI	Java	2	127 (LOC)	198
Yoo and Harman [42]	HC	C	4	NA	1267
Baudry et al. [9]	SI	Eiffel/C#	2	35 (classes)	1647
Ghiduk [7]	EO	Java	13	927 (LOC)	1772
Papadakis and Malevris [12]	HC	Java	8	395 (LOC)	2759
May et al. [21]	SI	Java	4	290 (LOC)	3958
Patrick et al. [45]	EO	Java	10	1945 (LOC)	8114
Fraser and Zeller [11]	EO	Java	10	1416 (classes)	117,913

NA not available

said to be an effective technique for automatically generating test suites to kill mutants.

Metaheuristic optimisation techniques are able to select test cases more efficiently than random testing because they generate test cases by taking into account the results of previous tests. This can be understood through the use of a simple example (see Fig. 8). Consider the problem of selecting test cases to reach two mutants (M1 and M2). Assume the three input variables (a, b and c) have integer input domains from 0 to 9 inclusive. The probability of a randomly selected test case reaching M1 is $0.1 * 0.9 * 0.1 = 0.009$ and the probability of it reaching M2 is $0.1 * 0.1 * 0.2 = 0.002$. Metaheuristic optimisation may measure the distance between the current and required value at each branch condition to target the correct branch.

If in our example the value of 'a' at the point of reaching the first branch condition is 6, the branch distance is measured as 1 (i.e. 6–5). The branch condition may thus be met by decreasing the value of 'a' by 1 (from 6 down to 5). Typically, the task of optimising input values to meet a particular branch condition is less trivial than this, since the value of a variable at a certain point in the program depends upon the branches and statements that were executed before. Nevertheless, metaheuristic optimisation techniques can use the information provided by metrics such as branch distance to guide the selection of test cases to reach and kill each mutant.

It is difficult to identify the most effective metaheuristic technique from the results of the empirical studies. The results cannot be directly compared because they use different test programs written in different program languages and with different settings for the techniques. Some studies compare multiple techniques, yet their conclusions are based upon experiments with only a small number of programs. Ayari claimed [15] ant colony optimisation was able to kill more than twice as many mutants as a genetic algorithm and more than three times as many as hill climbing. May [21] showed an artificial immune system to kill more mutants than a genetic algorithm with just 12.9 % the number of fitness evaluations. Similarly Baudry [9] showed a bacterial foraging algorithm to achieve 96 % mutation score in

Fig. 8 A simple branch structure leading to two mutants

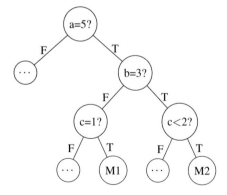

just 30 generations, compared to a genetic algorithm's 85 % after 200 generations. Yet in total, these three studies are based on experiments with just 8 programs. Certainly this is not enough to say swarm intelligence techniques are more effective in general. It is possible the researchers unwittingly chose programs for which it is more effective.

Each metaheuristic technique has advantages and disadvantages with respect to the program under test. For some programs, different parts of a test case can be evolved individually and then joined together by recombination in a genetic algorithm [39]. Yet for most programs, techniques such as hill climbing may be more efficient [40]. Similarly, there is no guarantee that techniques which reuse existing test cases (memetic algorithms and swarm intelligence) will be any more efficient than restarting the search. One suggestion is to evaluate a number of techniques in an attempt to determine empirically which one is the most effective for the program under test. However, in practice there is rarely enough time to fine-tune every technique to reach the full potential of its effectiveness. We need some general guidelines as to the techniques which might be effective for a particular program.

Hill climbing only stores one candidate solution at a time. This makes it difficult to take advantage of input values that were previously successful at killing mutants. Consider the example in Fig. 8. A test case may have been found to kill M1, but when targeting M2 the program could follow a different path. If this path includes a condition which prevents 'c' being less than two, this introduces a local optimum that hill climbing can only resolve by restarting the search. It may be possible to get around this problem by starting the search with input values from a test case that killed a nearby mutant. Yet, the steps taken from these values may still lead to a local optimum. The mutant might only be killed after a certain number of loop iterations or a particular branch is taken. Of course, hill climbing can be restarted as many times as is necessary, but in the worst case this degenerates to random testing.

Evolutionary optimisation techniques maintain a population of candidate solutions, which allows them to be more robust against local optima. Even if most of the candidates follow a path that prevents the goal being reached, a candidate which follows the right path can be used to create a new population and avoid having to restart the search. Techniques such as CMA-ES [45] use information about the rate at which fitness increases to maintain a population of candidates that is likely to be efficient at finding the global optimum solution. In addition, recombination can sometimes be applied to two locally optimum solutions to produce a candidate that is nearer to the global optimum. For example, when trying to kill M2, recombination can take the value of 'b' from the test case that fails at the 'c < 2' branch condition and the value of 'c' that kills M1. Since 'c = 1' is completely subsumed by the 'c < 2' condition, the new test case will kill M2. Evolutionary techniques are more likely to find the global optimum in program that as lots of local optima than hill climbing.

It seems counter-intuitive not to take advantage of local fitness gradients when available. Swarm intelligence combines the best of evolutionary optimisation and hill climbing in an approach that is population-based and involves local search.

They make it easier to kill M2, not only by remembering the test case which kills M1, but also a number of other tests that are found to be fit. Yet it is difficult to uniquely define these techniques, since evolutionary optimisation can also keep some of the fittest solutions in its population and variation may be applied locally as well as globally. Swarm intelligence has not yet been evaluated as thoroughly for test data generation as evolutionary optimisation. More work is required to determine what distinguishes these techniques, both in their definition and effectiveness.

6 Conclusions

Testing is a challenging, but vitally important part of the software development process. Time and effort spent on improving a test suite is worthwhile if it reduces the damage caused by faults and the work required to repair them later. Mutation Analysis may be used to evaluate the effectiveness of test cases and it provides indications as to how to improve them. The effectiveness of a test suite can be improved iteratively by combining mutation analysis with metaheuristic optimisation. Metaheuristic optimisation is highly suitable for this task because it makes few assumptions about the problem being solved. Rather than specifying how to produce an effective test suite, metaheuristic optimisation searches the fitness landscape iteratively and takes advantage of patterns of suitability as it goes along.

This chapter explored the application of three kinds of metaheuristic optimisation technique for test data generation. Hill climbing is efficient at fine tuning test cases, but depending on the starting conditions, it is prone to becoming stuck in a local optimum. Evolutionary optimisation is able to find the global optimum even when the fitness landscape is large, complex, noisy or discontinuous. However, swarm intelligence is more adaptable to change because it optimises a range of potential solutions and favours diversity over perfection. This means that when the software changes, or new unexpected faults are found, swarm intelligence is able to react more quickly to the new situation than other techniques. On the other hand, since swarm intelligence is based upon redundancy, it can be less efficient than other techniques and the candidate solutions it produces are not equally as effective. Each technique has its advantages and disadvantages, so it is necessary to choose between them according to purpose, or take a hybrid approach using multiple techniques.

Metaheuristic techniques are often applied as a 'black box' tool for optimisation. This has its obvious advantages, but it also brings with it some problems. First, the results can be difficult to understand. Why were these input values chosen? What is the purpose of this particular test case? Secondly, it is difficult to know what the best settings for a metaheuristic technique will be. How should we set the fitness function? Should we use recombination or elitism etc. and what should the termination criteria be? Finally, they can also take longer than deterministic techniques. We need to be prepared to leave metaheuristic optimisation techniques running in the background and allow enough time for them to produce an effective test suite.

The key to success when designing a metaheuristic technique for mutation analysis is not to focus too much on the metaphor that is being used, but on the decisions that cut across the different kinds of optimisation technique. A suitable balance must be found between intensifying and diversifying the search. A decision must also be made regarding the number of candidate solutions to maintain. A large population allows a broad range of options to be explored and is less likely to become stuck in a local optimum, but it tends to be less appropriate for fine-tuning the solution. The choice of fitness function is also very important. Good fitness functions must differentiate clearly between desirable and undesirable candidate solutions, but not be too expensive to calculate. It is not usually possible to make the correct decisions in advance, so it is important to compare and evaluate the different options empirically.

References

1. Talbi, E.-G.: Metaheuristics: From Design to Implementation. Wiley, Hoboken (2009)
2. Yang, X.-S.: Metaheuristic optimization: algorithm analysis and open problems. Lect. Notes Comp. Sci. **6630**, 21–32 (2011)
3. McMinn, P.: Search-based software test data generation: a survey. Softw. Test. Verif. Reliab. **14**, 105–156 (2004)
4. Bianchi, L., Dorigo, M., Gambardella, L.M., Gutjahr, W.J.: A Survey on metaheuristics for stochastic combinatorial optimization. Nat. Comput. **8**, 239–287 (2009)
5. Jia, Y., Harman, M.: An analysis and survey of the development of mutation testing. IEEE Trans. Softw. Eng. **37**, 649–678 (2011)
6. Offutt, A.J.: Investigations of the software testing coupling effect. ACM Trans. Softw. Eng. Methodol. **1**, 5–20 (1992)
7. Ghiduk, A.S.: Using evolutionary algorithms for higher-order mutation testing. Int. J. Comp. Sci. Issues **11**, 93–104 (2014)
8. Mishra, K.K., Tiwari, S., Kumar, A., Misra, A.K.: An approach for mutation testing using elitist genetic algorithm. In: Proceedings of IEEE International Conference Computer Science Information Technology 426–429 (2010)
9. Baudry, B., Fleurey, F., Jézéquel, J.-M., Le Traon, Y.: From genetic to bacteriological algorithms for mutation-based testing. Softw. Test. Verif. Reliab. **15**, 73–96 (2005)
10. Bottaci, L.: A genetic algorithm fitness function for mutation testing. In: Proceedings of International Works. Software Engineering Metaheuristic Innovative Algorithms, pp. 3–7 (2001)
11. Fraser, G., Zeller, A.: Mutation driven generation of unit tests and oracles. In: Proceedings of 19th International Symposium Software Testing Analysis, pp. 147–158 (2010)
12. Papadakis, M., Malevris, N.: Searching and generating test inputs for mutation testing. SpringerPlus **2** (2013)
13. Korel, B.: Automated software test data generation. IEEE Trans. Softw. Eng. **16**, 870–879 (1990)
14. Kaur, A., Goyal, S.: A bee colony optimization algorithm for fault coverage based regression test suite prioritization. Int. J. Adv. Sci. Tech. **29**, 17–30 (2011)
15. Ayari, K., Bouktif, S., Antoniol, G.: Automatic mutation test input data generation via ant colony. In: Proceedings of 9th Annual Conference Genetic Evolutionary Computation, pp. 1074–1081 (2007)
16. Myers, G.J., Sandler, C., Badgett, T.: The Art of Software Testing. Wiley, Hoboken (2012)

17. Blair, M., Obenski, S., Bridickas, P.: Patriot missile software problem. Technical report GAO/IMTEC-92-26, United States General Accounting Office (2002)
18. Heusser, M.: Software testing lessons learned from Knight Capital Fiasco. CIO Magazine (2012). http://www.cio.com/article/713628/Software_Testing_Lessons_Learned_From_Knight_Capital_Fiasco/. Cited 28 Sep 2014
19. Bezier, B.: Software Testing Techniques. Van Nostrand Reinhold, New York (1990)
20. Jay, F., Mayer, R.: IEEE standard glossary of software engineering terminology. Technical report 610.12-1990, IEEE (1990)
21. May, P.S.: Test data generation: two evolutionary approaches to mutation testing. Ph.D. thesis, Department of Computer Science, University of Kent, Canterbury, UK (2007)
22. Pacheco, C., Lahiri, S., Ball, T.: Finding errors in .NET with feedback-directed random testing. Technical report MSR-TR-2008-29, Microsoft Research (2008)
23. Sun Microsystems: Java Compatability Kit 6b User's Guide. Sun Microsystems (2012). http://openjdk.java.net/groups/conformance/docs/JCK6bUsersGuide/JCK6b_Users_Guide.pdf. Cited 28 Sep 2014
24. Oracle: Java Bug Database. Oracle (2012). http://bugs.sun.com. Cited 28 Sept 2014
25. Dijkstra, E.W.: Notes on structured programming. In: Dahl, O.J., Dijkstra, E.W., Hoare, C.A.R. (eds.) Structured Programming, pp. 1–82. Academic Press Ltd., London (1972)
26. Mahmood, S.: A systematic review of automated test data generation techniques. Masters thesis, School of Engineering., Institute of Technology Box, Ronneby, Sweden (2007)
27. Zhu, H., Hall, P.A.V., May, J.H.R.: Software unit test coverage and adequacy. ACM Comput. Surv. **29**, 366–427 (1997)
28. Lipton, R.: Fault diagnosis of computer programs. Technical report, School of Computer Science, Carnegie Mellon University (1971)
29. Jia, Y., Harman, M.: Java mutation testing repository. UCL (2014). http://crestweb.cs.ucl.ac.uk/resources/mutation_testing_repository/. Cited 28 Sept 2014
30. Budd, T.A.: Mutation analysis of program test data. Ph.D. thesis, Department of Computer Science, Yale University, New Haven, CT (1980)
31. DeMillo, R.A., Lipton, R.J., Sayward, F.G.: Hints on test data selection: help for the practicing programmer. Computer **11**, 34–41 (1978)
32. Offutt, A.J., Untch, R.H.: Mutation 2000: uniting the orthogonal. In: Wong, W.E. (ed.) Mutation Testing for the New Century, pp. 34–44. Kluwer Academic Publishers, Norwell (2001)
33. Blum, C., Roli, A.: Metaheuristics in combinatorial optimization: overview and conceptual comparison. ACM Comput. Surv. **35**, 268–308 (2003)
34. Sörensen, K.: Metaheuristics—the metaphor exposed. Int. Trans. Oper. Res. (in press)
35. Burnstein, I.: Practical Software Testing: A Process-Oriented Approach. Springer, New York (2003)
36. Mala, D.J., Mohan, V.: Quality improvement and optimization of test cases: a hybrid genetic algorithm based approach. ACM SIGSOFT Softw. Eng. Notes **35**, 1–14 (2010)
37. Burke, E.K., Kendall, G.: Search methodologies: introductory tutorials in optimization and decision support techniques. Springer, New York (2010)
38. Simon, D.: Evolutionary Optimization Algorithms. Wiley, Hoboken (2013)
39. Forrest, S., Mitchell, M.: Relative building-block fitness and the building-block hypothesis. In: Whitley, D. (ed.) Foundations of Genetic Algorithms 2, pp. 109–126. Morgan Kaufmann, San Mateo (1993)
40. Harman, M., McMinn, P.: A theoretical and empirical analysis of evolutionary testing and hill climbing for structural test data generation. In: Proceedings of 16th International Symposium Software Testing Analysis, pp. 73–83 (2007)
41. Arcuri, A.: Full theoretical runtime analysis of alternating variable method on the triangle classification problem. In: Proceedings 1st International Symposium Search Based Software Engineering, pp. 113–121 (2009)
42. Yoo, S., Harman, M.: Test data regeneration: generating new test data from existing test data. Softw. Test. Verif. Reliab. **22**, 171–201 (2012)

43. Eiben, A.E., Smith, J.E.: Introduction to Evolutionary Computing. Springer, New York (2003)
44. Koza, J.R.: Genetic Programming III: Darwinian Invention and Problem Solving. Morgan Kaufmann, San Francisco (1999)
45. Patrick, M., Alexander, R., Oriol, M., Clark J.A.: Selecting highly efficient sets of subdomains for mutation adequacy. In: Proceedings 20th Asia-Pacific Software Engineering Conference, pp. 91–98 (2013)
46. Bonabeau, E., Dorigo, M., Theraulaz, G.: Swarm Intelligence: From Natural to Artificial Systems. Oxford University Press, New York (1999)
47. Blum, C., Merkle, D.: Swarm Intelligence: Introduction and Applications. Springer, New York (2008)
48. Camazine, S., Deneubourg, J.L., Franks, N.R., Sneyd, J., Theraula, G., Bonabeau, E.: Self-organization in Biological Systems. Princeton University Press, Princeton (2003)
49. Serugendo, G.M., Gleizes, M.-P., Karageorgos, A.: Self-organising Software: From Natural to Artificial Adaptation. Springer, New York (2011)
50. May, P., Timmis, J., Mander, K.: Immune and evolutionary approaches to software mutation testing. Lect. Notes Comput. Sci. **4628**, 336–347 (2007)
51. Passino, K.M.: Biomimicry of bacterial foraging for distributed optimization and control. IEEE Control Syst. **22**, 52–67 (2002)
52. Papadakis, M., Malevris, N.: Automatic mutation test case generation via dynamic symbolic execution. In: Proceedings of 21st International Symposium Software Reliability Engineering, pp. 121–130 (2010)
53. Harman, M., Yue, J., Langdon, W.B.: Strong higher order mutation-based test data generation. In: Proceedings of 21st ACM SIGSOFT Symposium Foundations Software Engineering, pp. 212–222 (2011)
54. Chen, T.Y., Kuo, F.-C., Liu, H., Wong, W.E.: Code coverage of adaptive random testing. IEEE Control Syst. **62**, 226–237 (2013)

Author Biography

Matthew Patrick is a Research Associate in the Department of Plant Sciences, in the University of Cambridge. He is currently working on the characterisation and regeneration of heterogeneous host landscapes for epidemiological modelling, although his interests also include software testing and search-based software engineering. Matthew strives to work at the confluence between Biology and Software Engineering.

Measuring the Utility of Functional-Based Software Using Centroid-Adjusted Class Labelling

Nick J. Pizzi

Abstract The functional programming paradigm involves stateless computation on immutable data constructs. While this paradigm's historical context dates back to the early twentieth century with lambda calculus and a formal study of computability and function definition, there has been a resurgence in functional programming, especially in the area of predictive analytics. New, purely functional, languages have recently emerged, and functional extensions have been added to several popular programming languages. It is sometimes difficult to estimate the overall utility and extensibility of functional programming software components. At the same time, many software metrics exist that attempt to quantify various qualitative attributes of software components. Here, we use a computational intelligence strategy that uses a set of software metrics to predict the qualitative utility of a software system's underlying components. Centroid-adjusted class labelling is a pattern classification preprocessing method that compensates for the possible imprecision of an established external reference test (gold standard) by adjusting, when necessary, design pattern class labels while maintaining the reference test's discriminatory power. The adjusted design labels incorporate within-class centroid information using robust measures of location and dispersion. This method is applied to a biomedical data analysis software system written in a functional programming style. It is shown that significant improvement to the discriminatory power of the classifier is obtained when using this preprocessing method.

Keywords Functional programming · Pattern classification · Software utility · Software metrics · Java lambda · Software engineering

N.J. Pizzi (✉)
InfoMagnetics Technologies Corporation, Research and Technology Development,
Winnipeg, MB R3C 3Z5, Canada
e-mail: pizzi@imt.ca

N.J. Pizzi
Department of Computer Science, University of Manitoba, Winnipeg, MB R3T 2N2, Canada

© Springer International Publishing Switzerland 2016
W. Pedrycz et al. (eds.), *Computational Intelligence and Quantitative Software Engineering*, Studies in Computational Intelligence 617,
DOI 10.1007/978-3-319-25964-2_6

1 Introduction

Software systems are utilized to model increasingly sophisticated problems across many application domains. Given the complexity of these contemporary systems, it is often difficult to gauge the utility of their underlying software components. A possible strategy to evaluate the qualitative attributes of a system's components is to use software metrics as quantitative predictors. Examples of software metrics include the coupling between components, the number of lines of code, the number of unique operators and operands, cyclomatic complexity, and the ratio of lines of comments to code. If an external reference test can be identified to label the utility of these components then this strategy may be viewed as a problem of classification. That is, predict a component's qualitative attribute (based on the reference test) using a set of metrics. No single metric can be an indicator of quality for all types of software systems; however, determining the "optimal" combination of multiple software metrics would certainly be a non-trivial exercise. Incorporating the expertise of software architects into a "quality filter" analysis system may be an ideal complement to finding a mapping from metrics to quality. Couching the mapping problem as one of classification where the qualitative assessments are initially determined by the expert is a potentially viable option as will be demonstrated in this investigation.

Pattern classification involves finding a mapping (relationship) from patterns to a set of classes. Patterns are composed of features (for example, a set of software metrics) and class labels (for example, software component utility) are assigned using an external reference test (for example, assessment by a software architect). Effective pattern classification requires the coupling of an efficient classifier with a synergistic preprocessing strategy. The motivation for pre-processing strategies [42, 68] is to simplify the determination and construction of class boundaries in the feature space. For instance, Grandvalet and Canu [25] describe the identification of relevant input variables through a constrained scaling strategy for a support vector machine [73] used in a facial expression recognition task. A sequential search method is presented in Pudil et al. [57], which iteratively examines varying numbers of features, and is shown to be more computationally efficient than a branch and bound method. Using a set of multispectral images in a prostate cancer classification task, Tahir et al. [67] empirically demonstrate the effectiveness of feature selection using a hybrid classifier approach combining the K-nearest neighbour algorithm [12] with a search heuristic based on Tabu search [24]. Schmitt et al. [59] empirically evaluate a feature extraction strategy employing Choquet integration [48] within a fuzzy rule classifier, and demonstrate that by extracting relevant features the total number of fuzzy rules generated is reduced. Preprocessing strategies also exist to deal with the possible imprecision of external reference tests while maintaining their essential (and domain-accepted) discriminatory power [52, 53, 71, 76].

While reference tests may be well-established benchmarks, they are seldom perfectly accurate and sometimes incorrectly applied. Nevertheless, any method that compensates for their imprecision must ensure that mappings are correctly validated against these benchmarks. Factors contributing to a tarnished gold standard include subjective estimates by a domain expert (or expert panel), simple clerical errors, unreliable or imperfect sample acquisition techniques, or anomalous sensor readings. One such preprocessing method, *centroid-adjusted class labelling* [54], compensates for the possible imprecision of class labels using a fuzzy similarity measure based on robust measures of location and dispersion for each class of design patterns. In essence, this produces, for each pattern, a list of "membership" values (on the unit interval) for every class. If a design pattern is sufficiently dissimilar to other patterns from its class and sufficiently similar to those from another class, then its class label is changed to the latter class.

We empirically evaluate the centroid-adjusted class labelling preprocessing method using software components from a biomedical data analysis software system implemented using the functional programming paradigm. The patterns comprise software metrics for each component and the class labels are measures of software utility as determined by a domain expert (a software architect). We begin with a discussion of the functional programming paradigm in Sect. 2. Section 3 describes pattern classification and performance assessment and Sect. 5 provides details of the centroid-adjusted class labelling method. The experiment design and results are presented in Sects. 6 and 7, respectively, followed by some concluding remarks.

2   Functional Programming

Functional programming is a paradigm that treats computation in terms of immutable values and functions that translate between them [22]. The notion of functional programming goes back to the 1930s with Church's [10] formulation of the lambda calculus, a formal mathematical logic system for the expression of computation based on function abstraction and application, recursion, and variable binding and substitution. A key difference between functional programming and other paradigms is the notion of referential transparency. With functional programming, the output value of a function depends only on its input; eliminating "side effects" can make it easier to understand and predict program behaviour, hence one of the main motivators in its adoption. While the functional paradigm has methodological benefits [3, 33, 34, 50], functional languages have been slower than their imperative and object-oriented counterparts, but with recent compiler optimizations and well-designed data structures this gap is much narrower and much less of a concern. Functional programming languages include Haskell [49], Clojure [17], Erlang [8], and Objective Caml [40]. Programming in a functional style can also be accomplished in some imperative and object-oriented through functional extensions: C# [66], F# [64], R [1], Mathematica [44] and Java 8 [78]. As the biomedical data analysis system used in the experiment described in Sect. 7 was written in Java 8 using its new lambda expressions, we now discuss some of the functional programming extensions for this most recent version of Java.

2.1 Functions and Java 8

In Java 8, functions are treated very differently from methods. For instance, functions are independent of any component instance, they do not access global state, modify input, or change state; that is, they possess complete referential transparency. Functions may be stored in variables, passed as parameters, or returned from other functions. Let us look at several examples. First, here is a lambda expression involved in updating a time series value based on the current time step:

```
IntToDoubleFunction newMagnitude = timeStep -> 2.28 +
8.87 * timeStep;
```

where `IntToDoubleFunction` represents a function that accepts an integer-valued argument and produces a double-valued result, `timeStep` is the input, and `newMagnitude` is a function. Note that this is similar in notation to a Java anonymous implementation of an interface with a single method. We can also use a generic function interface for a function from one object to another, for example,

```
Function<String, Integer> geneCount = s-> s.split(" ").length;
```

where the first type, `String`, is the parameter and the second type, `Integer`, is the return type. We can also make lambda expressions with multiple parameters, in which case parentheses are needed around the parameter names. They may also contain more than one expression, in which case curly braces are needed around the expressions as well as a "return" for the last expression, as in

```
BiFunction<String, Integer, Boolean> exceedsMaxLen = (s,
maxLen) -> {

int actualLen = s.length();
return actualLen > maxLen;

};
```

where `BiFunction` represents a function that accepts two arguments and produces a result (a two-arity specialization of `Function`).

3 Pattern Classification

3.1 Design and Validation

We now introduce some formal notation to more accurately describe the design and validation of a pattern classifier. A classification exercise always involves a set of N pattern-class pairs, $\mathbf{X}$

$$\mathbf{X} = \{(\mathbf{x}_k, w_k), (k = 1, 2, \ldots, N) | \mathbf{x}_k \in \mathfrak{R}^n, w_k \in \mathbf{w} = \{1, 2, \ldots, c\}\} \qquad (1)$$

where $\mathbf{x}_k = [x_1, x_2, \ldots, x_n]$ is an n-dimensional *pattern* and w_k is its corresponding *class*. Let us divide the pattern space into a set of $\mathbf{W} = \{W_1, W_2, \ldots, W_c\}$ regions. Now, if $\mathbf{x}_k$ is found in the region W_j, then $w_k = j$. Pattern classification is a two-stage empirical process: *design* and *verification*. Design is the process of finding a mapping (decision boundaries), f, which partitions the patterns into k spatial regions, such that, if a pattern has the class label, w_j, it will also lie in the spatial region, W_j, containing all and only those patterns with class label, w_j. On the other hand, verification refers to the application of the mapping, found during the design process, to new (previously unseen) patterns in order to predict their respective class labels. Finally, let us define the *feature*, p_i, as the set of x_i values for all patterns, $\mathbf{x}$.

In our specific software engineering investigation: the features are the software metrics that are used in the experiment; x_j is the set of values for the respective metrics for the jth software component; and w_j is the utility level (class) for component j. The utility level (for instance, *low*, *medium*, and *high*) would be determined by an external reference test (for instance, a software engineer or panel of developers).

An important question to address during the design process is the performance of the classifier, or, in other words, the accuracy of the mapping [47]. Prior to selecting the performance measure, we randomly allocate the patterns into two subsets: N^D design patterns, $\mathbf{X}^D$; and N^V validation patterns, $\mathbf{X}^V$ ($N^D + N^V = N$). The design subset is used to find the mapping, $f: \mathbf{X}^D \rightarrow \mathbf{w}$. The validation subset is used to evaluate the accuracy (performance) of the decision rule, $f: \mathbf{X}^V \rightarrow \mathbf{w}$. For example, say (x_j, w_j) is a pair where $x_j \in \mathbf{X}^V$ and f produces the class, w_j'. Now, if $w_j' = w_j$, f has correctly predicted the class for x_j, otherwise f has made an incorrect prediction (a classification error).

3.2 Performance Assessment

One issue that needs to be addressed is how the performance of a classifier should be assessed using the patterns from the validation subset. A typical approach is to construct a c × c accuracy matrix, $\mathbf{A} = [a_{ij}]$, of predicted versus assigned classes

[14]. Say, we have a validation pattern assigned to class i and the classifier mapping predicted that the pattern belonged to class j: if i = j, then a_{ij} is incremented by one, otherwise a misclassification occurs and a_{ij} is incremented. The conventional performance measure used in much of the pattern classification literature is P_O, the ratio of correctly classified validation patterns to N^V

$$P_O = N_V^{-1} \sum_i a_{ii} (i = 1,2,\ldots,c) \tag{2}$$

Unfortunately, P_O does not take into account any agreement that may be due to chance, P_L

$$P_L = N_V^{-2} \sum_i \left(\sum_j a_{ij} \sum_j a_{ji} \right) (i,j = 1,2,\ldots,c) \tag{3}$$

A more conservative measure is the κ score [18, 21], a chance-corrected measure of agreement between the predicted and assigned classes

$$\kappa = \frac{P_o - P_L}{1 - P_L} \tag{4}$$

where κ = 1 means there is perfect agreement between the predicted and assigned classes, κ = 0 means that the agreement is due strictly to chance, κ > 0 means that there is partial agreement not due to chance, and, finally, κ < 0 means that the agreement is actually less than chance (with the floor depending upon the marginal distributions of the validation patterns within each class [39]).

3.3 Classifiers

Myriad classifiers, algorithmic systems used to construct decision boundaries, exist with different properties, architectures, advantages and limitations. Details of these various considerations may be found in [2, 4, 14, 15, 16, 28, 47, 69]. We now briefly describe the pattern classifier that is used in this investigation.

The support vector machine (SVM) [60, 74] is an important family of supervised learning algorithms that are used for pattern classification. SVM selects class models that maximize the error margin of a design subset. For instance, given a set of patterns that belong to one of two classes, an SVM finds the hyperplane leaving the largest possible fraction of patterns of the same class on the same side while maximizing the distance of either class from the hyperplane. The approach is usually formulated as a constrained optimization problem and solved using constrained quadratic programming. While the original method [75] could only produce linearly separable class regions, it may be extended by introducing non-linear "kernels" [77] that exploits the fact that a nonlinear mapping of sufficiently high dimension can project the patterns to a new parameter space in which the classes may be separated by a hyperplane. In general, it cannot be determined a priori which kernel will contribute to producing the best classification results for a given dataset and one must rely on trial and error experimentation. For patterns x_i and x_j, common kernel functions (a, b, and d are user-defined scalars), $K(x_i, x_j)$, are: power, $(x_i \cdot x_j)^d$; polynomial, $(a\, x_i \cdot x_j + b)^d$; sigmoid, $\tanh(a\, x_i \cdot x_j + b)$; and Gaussian, $\exp(-0.5|\, x_i - x_j|2/\sigma)$. In the interest of brevity, we restrict ourselves to the Gaussian kernel for this investigation.

4   Software Attributes and Measuring Utility

A software metric [6, 9, 36–38, 51, 79] is a mapping from a software component (object, function, method, class, and so on) to a set of scalar values to quantify one or more qualitative attributes [19]. Metrics are commonly regarded as important factors reflecting the nature of software systems including characteristics such as the qualitative properties of complexity, extensibility, and, in our case, utility [55, 58, 62]. The proper use of metrics described in the software engineering literature [20, 31, 43, 56, 65] requires a fundamental understanding of what is being measured and a sound interpretation of how the measurements are being acquired both of which are necessary prerequisites to refactoring, that is, changing a software component to improve its conceptual structure without affecting its behaviour [23, 45]. Metrics may be used by developers to identify "high risk" software components; that is, components that are too complex to understand and test, exhibit poor usability for clients, or are not sufficiently reusable. Maintainability, as defined by the ISO/IEC 9126 standard [36], is the set of attributes that bear on the effort needed to make specified modifications, and includes the notions of stability, analyzability, changeability, and testability. It also involves the concept of how difficult it is to make small or incremental changes to existing software components without introducing errors in logic or design. Utility involves the amount and quality of code comments, the overall size of the component, and the overall flow of control [7, 36].

This investigation did not involve an a priori determination that the software metrics used here were, in some sense, optimal or ideal predictors of software component utility. It is more than likely that this set is neither wholly necessary nor sufficient for this purpose. These metrics were selected for the purely practical reason that metric generation software was available to compile this specific set of metrics.

4.1 Metrics Used as Features

Table 1 lists the software metrics, and their corresponding feature indices, used in the experiments described in Sect. 6. Several procedural metrics are used: the number of lines of code, p_1; the ratio of comments to lines of code, p_2, and comments to non-comments, p_3; the total number of operands and operators, p_4 and p_5; the total number of unique operands and operators, p_6 and p_7; the number of functions, p_{23}; the maximum number of parameters defined for a single operation, p_8; the nesting depth of decision/loop constructs; the number of attributes, p_9; the number of reference types used in attribute declarations, p_{10}. McCabe's [46] cyclomatic complexity, p_{11}, of a component is the number of possible paths in its decision flow graph (informally, one more than the number of binary decisions) and p_{24} is the number of functions weighted by p_{11}.

While functional programming is a paradigm different from object-oriented programming, some of the latter's metrics [9, 29] are relevant in functional software code, especially for hybridized languages such as Java 8 [78]. (Of course, some are not relevant such as metrics relating to inheritance; however, even in object-oriented programming the notion of inheritance is being replaced by inter-face design methods [11].) Coupling between components, p_{12}, is the count of reference types that are used in attribute declarations, formal parameters and return types, and represents the number of other components to which component is coupled (if this metric is large, modularity decreases). Abstraction coupling, p_{13}, is the number of reference types used in attribute declarations. Another potentially

Table 1 Summary of software metrics

Feature	Metric	Feature	Metric
p_1	Lines of code	p_{15}	Number of members
p_2	Comment ratio 1	p_{16}	Program length
p_3	Comment ratio 2	p_{17}	Program vocabulary
p_4	Number of operands	p_{18}	Program volume
p_5	Number of operators	p_{19}	Program difficulty
p_6	Number of unique operands	p_{20}	Program effort
p_7	Number of unique operators	p_{21}	Attribute Access
p_8	Maximum number of parameters defined for a single operation	p_{22}	Function dissimilarity
p_9	Nesting depth of decision/loop constructs	p_{23}	Number of functions
P_{10}	Number of attributes	p_{24}	Number of functions weighted by cyclomatic complexity
p_{11}	Cyclomatic complexity	p_{25}	Call-graph fan-in
p_{12}	Component coupling	p_{26}	Call-graph fan-out
p_{13}	Abstraction coupling	p_{27}	Call-graph nesting depth
p_{14}	Acquaintance components	p_{28}	Call-graph edges

useful metric is the number of "acquaintance" components, p_{14}, for a given component [41]; large values for this metric may suggest excess coupling, which is detrimental to software reuse. We also include a component's total number of members, p_{15}, the percentage of a component's functions that do not access a particular attribute averaged over all of its attributes, p_{21}, and $p_{22} = (\Sigma_j \, r_j - r)/(s - s \times r)$ (where r is the number of methods, s is the number of attributes, and r_j is the number of functions that access attribute j) measures the dissimilarity of functions in a component by attributes [29].

While some controversy exists concerning their utility [35], we also use software metrics introduced by Halstead [27] (as they are used in software engineering to measure complexity), including: program length, $p_{16} = p_4 + p_5$; program vocabulary, $p_{17} = p_6 + p_7$; program volume, $p_{18} = p_{16} \times \log_2 p_{17}$; program difficulty, $p_{19} = (p_7/2) \times (p_4/p_6)$; AND, program effort, $p_{20} = p_{18} \times p_{19}$.

Finally, we use several graph-based metrics that are germane to functional programming including fan-in, p_{25}, fan-out, p_{26}, nesting depth, p_{27}, and number of edges, p_{28}. Van den Berg and van den Broek [72] describes in detail the use of these metrics with respect to function-based call-graphs.

5 Adjusting Design Class Labels

Class labels assigned by an external reference test are not necessarily perfect due to issues relating to subjective estimates by a subject matter expert (or panel of experts), clerical errors, unreliable sample acquisition techniques, or anomalous sensor readings. A useful pre-processing strategy, prior to pattern classification, is to compensate for the possible imprecision of the assigned design classes while simultaneously maintaining their domain accepted discriminatory power. There are three approaches to design class adjustments for the purpose of mitigating the effects of potentially unreliable reference tests [54]: (i) *reassignment*, a pattern's class is changed if it is more "similar" to patterns from another class; (ii) *surrogation*, introduces new classes $\mathbf{w}^S = \{1, 2, ..., d\}$; and (iii) *gradation*, patterns belong to all classes to varying degrees, thereby moving from a crisp assignment, $\mathbf{w} = \{1, 2, ..., c\}$, to a fuzzy [81] assignment, $\mathbf{w}^G = [0,1]^c$.

One such pre-processing strategy, which uses reassignment, is centroid-adjusted class labeling (CACL) [53] that compensates for potential class assignment imprecision by using a similarity measure based on robust measures of location and dispersion. If a design pattern is sufficiently dissimilar to other design patterns from its class and sufficiently similar to patterns from another class then its class is reassigned to the latter one. The centre of each class (centroid) is computed using robust measures of location (see below), as they are insensitive to slight deviations from requisite normality assumptions about the underlying pattern distribution [30, 32, 61]. The centroids are computed using only design patterns; hence, the efficacy of this method is always measured against the original external reference test. Dispersion-adjusted distances are computing between each design pattern and each class centroid. The further a design pattern is from a class centroid, the lower its membership value for that class; however, reassignment will only take place if the pattern is sufficiently distant from the centroid of its assigned class and sufficiently near the centroid of another class.

5.1   Robust Location and Dispersion Measures

The breakdown point, $b \in [0.0, 0.5]$, which measures the robustness of location estimators, indicates what proportion of the data may be contaminated without significantly affecting the value of the estimator [13]. For example, $b \to 0$ for the mean since any one extreme value can cause this location estimator to become correspondingly large. An affine equivariant location estimator is unaffected by changes in shift and scale. So, a good location measure should have the properties of affine equivariance and $b \to 0.5$.

The median is an excellent location estimator for the univariate case precisely because it is affine equivariant and has a high breakdown point, $b = 0.5$. For the multivariate case, the feature-wise median [26], $\mathbf{m}^c = [z_1, z_2, \ldots, z_n]$ (where z_i is the median of feature p_i), has the same high breakdown point as the univariate median but, while it is shift and scale invariant, it is not orthogonally invariant. The spatial median [5], $\mathbf{m}^s$, minimizes

$$\sum_{i=1}^{N} \|x_i - \mathbf{m}^s\| \tag{5}$$

where $\|.\|$ is the Euclidean distance. This median has a high breakdown point, $b \approx (N + 1)/2N$, is shift and orthogonally invariant, but it is not scale invariant.

For this investigation, we use the halfspace median [70], $\mathbf{m}^h$, which is affine equivariant and has a reasonably high breakdown point, $b = [1/(N + 1), 1/3]$. The halfspace median is defined as the set of points, $\{\Theta\}$, of maximal depth, where a pattern's location depth, Θ, relative to a dataset is defined as the smallest number of patterns in a closed halfspace with boundary through Θ. Generally, there is no unique point; however, Small [63] has shown that this set of points is closed, bounded, and convex. Since, CACL requires only a single median, we will randomly select a single point in $\{\Theta\}$, if there is more than one.

We define a robust measure of dispersion for the class k centroid, τ_k, as

$$\tau_k(x) = \frac{\mathbf{m}^h(|x - \mathbf{m}^h(x)|)}{v} \tag{6}$$

where $\mathbf{x}$ are class k patterns and v is set to 0.6745 (similar to the univariate median of absolute deviations, this ensures that τ_k behaves as the standard deviation when the error distribution is normal [80]).

We may now define a similarity measure (inverse weighted distance), s_k, between a design pattern, $\mathbf{x}_j$, and the class k centroid $\mathbf{m}_k^h$

$$s_k(\mathbf{x}_j) = \left(1 + \left|\frac{\mathbf{x}_j - \mathbf{m}_k^h}{\tau_k}\right|\right)^{-1} \tag{7}$$

The set, $\mathbf{s}^j = \{s_1(\mathbf{x}_j), s_2(\mathbf{x}_j), \ldots, s_c(\mathbf{x}_j)\}$, represents the similarity of $\mathbf{x}_j$ to each class centroid. In other words, $\mathbf{x}_j$ can be said to belong to each class to varying degrees; the further $\mathbf{x}_j$ is from the class k centroid, the smaller s_k (its "membership" value). Normally, the maximum value of $\mathbf{s}^j$ will correspond to the design pattern's assigned class, that is, the pattern's assigned class accurately reflects the "similarity" (or proximity) to its class centroid. However, if the maximum element in $\mathbf{s}^j$ corresponds to a class different from the one $\mathbf{x}_j$ was originally assigned then a class reassignment takes place. In the unlikely scenarios of two or more maximal elements in $\mathbf{s}^j$: (i) if the assigned class is one of the elements then no reassignment is made; (ii) if the assigned class is not one of the maximal elements but one of the other maximal elements is nearer to the assigned class, in some domain-specific sense (as determined by the reference test), then this class is chosen for reassignment; (iii) otherwise, the reassignment is either randomly sampled from the maximal values or $\mathbf{x}_j$ is excluded from the experiment.

Figure 1 diagrams the rationale behind this approach. We have a set of design patterns (n = 2, c = 2) with groupings (dark grey versus light grey) of class 1 and class 2 patterns and their respective spatial regions, W_1 and W_2, centroids $\mathbf{m}_1^h$ and $\mathbf{m}_2^h$, and dispersions, τ_1 and τ_2. Now, examine the patterns, $\mathbf{x}_i$ (small light grey circle) and $\mathbf{x}_j$ (small gradient grey), which belong to class 2. While $\mathbf{x}_i$ is outside the

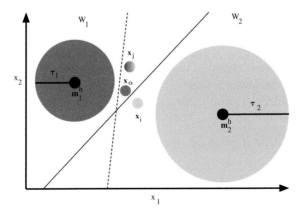

Fig. 1 Two classes of design patterns with respective centroids, $\mathbf{m}_1^h$ and $\mathbf{m}_2^h$, dispersions, τ_1 and τ_2, and spatial regions, W_1 and W_2. If the class 2 pattern, $\mathbf{x}_j$, is not reassigned then a classifier may produce a class boundary represented by the *dashed line*. However, if it is reassigned to class 1 then the same classifier may produce a potentially more representative boundary (*solid line*)

two-dimensional hypersphere of class 2 patterns, that is, $s_2(\mathbf{x}_i)$ is small, it is still greater than $s_1(\mathbf{x}_i)$ so no reassignment takes place (it is not sufficiently near $\mathbf{m}_1^h$). However, $\mathbf{x}_j$ is another matter; in this case, $s_1(\mathbf{x}_j) > s_2(\mathbf{x}_j)$. Hence, we reassign $\mathbf{x}_j$ from class 2 to class 1. What effect does this reassignment have with a classifier during the design process? With this reassignment, the class boundary is represented as the solid line in Fig. 1. Without the reassignment, the decision boundary is represented as the dashed line. Now, let us look at the validation process and, specifically, the validation pattern $\mathbf{x}_\alpha$ (smaller dark grey), with an assigned class of 1. With the reassignment made during the design phase, the validation pattern will be correctly classified, as it will lie in the W_1 spatial region; however, without the adjustment, $\mathbf{x}_\alpha$ will be misclassified, that is, the classifier mapping will assign it to class 2 as it lies in the W_2 spatial region.

6 Experiment Design

6.1 Software System

In this section, we present a brief discussion of the software whose components were extracted and assessed for the pattern classification experiment. The software is a biomedical data analysis system used to analyze stages of a disease relating to a hemostatic defect. This system interprets, visualizes, and aggregates information from a range of categories including laboratory results, responses from a detailed questionnaire, demographics, family history of illnesses, health risk factors, and various social determinants. Some specific features include: adenosine diphosphate; archidonic acid; bilirubin; alkaline phosphate; several blood cofactors (components with which others must unite in order to function), such as FVIIIc; age; underlying chronic conditions (for example, heart disease and asthma); screening records; immunizations; blood type; allergies; developmental history; and prescribed medications. This software system is written in Java using functional extensions, via lambda expressions, found in Java 8.

6.2 Quality Assessment by Architect

A software architect was asked to carefully scrutinize the software components of the biomedical data analysis system. Based on the architect's expert judgment, each component was assigned to one of three utility classes: low, medium, or high. The expert considered low quality components inferior, either in design or implementation; these components needed to be reviewed and possibly rewritten. The expert considered high quality components as well designed and well implemented; these components struck an excellent balance between functionality, performance, complexity, and developer ease of use. All other components were assigned a *medium* quality class label; these components were adequately designed and implemented, but not as well as *high* quality components, and did not require extensive review or revision, unlike the *low* quality components. The expert's subjective assessment was based on a number of software engineering considerations including: clarity of purpose; documentation adequacy; error handling and reporting; and overall complexity with respect to distribution of functions and attributes (this is an overall subjective evaluation and not the objective use of metrics described in Sect. 4); number of developers (found in comments) and their level of expertise; use of enumeration, typedef, and numeric constants; use of developer tags such as "todo", "future", and "implement"; the use of deprecated code; presence of code that has been commented out; the use of components that instantiate common software design patterns. After this thorough assessment, the expert assigned each component to one of the *high*, *medium*, or *low* classes.

7 Results and Discussion

The software component dataset ($N = 302$, $n = 28$, $c = 3$) was assessed by the software architect and was assigned to one of three utility classes: *low* ($N_1 = 75$), *medium* ($N_2 = 129$), and *high* ($N_3 = 98$). For the experiments, we use SVM with the Gaussian kernel with and without CACL preprocessing. We randomly allocate 50 patterns from each class (Table 2) to the design subset ($N^D = 1373$) with the remaining patterns assigned to the validation subset ($N^V = 684$).

During the design phase of the classification process, we present the design patterns with the original (non-adjusted) class labels to the SVM classifier. We then preprocess the patterns using CALC and present the same patterns, with potentially adjusted class labels, to the same classifier. Subsequently, for the validation phase, we use the mappings and the validation patterns to test the mappings' effectiveness and record the accuracy matrices and performance measures. Table 3 lists the number of adjustments made using CALC. A total of 17 (11 %) design patterns had their class labels altered. In the case of the *medium* utility class, for instance, three design patterns were reassigned to the *low* utility class and one to the *high* utility class, while five design patterns were reassigned to it (two from *low* and three from *high*). Interestingly, no design patterns were reassigned across two utility classes (*low* to *high* or *high* to *low*), which would be considered far apart in a domain (software engineering) sense.

Table 2 The random allocation of patterns into design, X^D, and validation, X^V, subsets

Class	N_i	Design	Validation
Low	$N_1 = 75$	50	25
Medium	$N_2 = 129$	50	79
High	$N_3 = 98$	50	48
Total	302	$N^D = 150$	$N^V = 152$

Table 3 The original and adjusted (using CACL) distributions of X^D patterns: y $\rightarrow$ z indicates that y design patterns were relabeled from their original utility class to class z, while y $\leftarrow$ z indicates that y design patterns were relabeled to class z from their original utility

Class	Original (N^D = 150)	CALC ($N^{D'}$ = 150)	Design adjustments
Low	50	51	2 $\rightarrow$ medium (3 $\leftarrow$ medium)
Medium	50	51	3 $\rightarrow$ low, 1 $\rightarrow$ high (2 $\leftarrow$ low, 3 $\leftarrow$ high)
High	50	48	3 $\rightarrow$ medium, (1 $\leftarrow$ medium)

Table 4 Classification accuracy matrix for SVM (i) and SVM with CACL (ii)

	Assigned versus predicted	Low	Medium	High	Performance
(i)	Low	57	14	4	κ = 0.63
	Medium	18	97	14	(P_O = 0.76)
	High	20	4	74	
(ii)	Low	64	10	1	κ = 0.78
	Medium	14	110	5	(P_O = 0.85)
	High	1	13	84	

Table 4(i) lists the classification accuracy matrix for SVM (with the Gaussian kernel) using the validation patterns without CACL. We see that SVM had an overall classification accuracy of κ = 0.63 (P_O = 0.76) and 8 % of the validation patterns were incorrectly predicted to belong to a utility class that was far (in a domain sense) from the assigned class (4 *low* patterns were predicted to belong to the *high* utility class and 20 *high* patterns were predicted to belong to the *low* utility class).

Table 4(ii) lists the classification accuracy matrices for SVM (with the Gaussian kernel) using the validation patterns and CACL. The overall classification accuracy is κ = 0.78, which is a 24 % improvement over SVM without CACL preprocessing (there is also a 12 % improvement with respect to P_O (0.85 versus 0.76)). Only 1 % of the validation patterns were incorrectly predicted to belong to a utility class that was far (in a domain sense) from the assigned class (1 *low* patterns were predicted to belong to the *high* utility class and 1 *high* patterns were predicted to belong to the *low* utility class).

After the classification experiments were completed and the performance results tabulated, we applied CACL to the validation patterns to ascertain if any of the validation patterns would have had their labels adjusted had they instead been included in the design subset. A total of 7 patterns would have had their class labels changed: 3 *low* patterns would have become *medium* patterns; and 3 *medium* patterns would have become *high* patterns; and 1 *high* pattern would have become a *medium* pattern. It is an interesting aside to examine what would have happened with the classifier performance if we had actually adjusted the class labels of these 7 validation patterns. Using the same mappings that produced the results in Table 4, we would have had better performance for the SVM classifier, $\kappa = 0.81$ ($P_O = 0.88$). Given this post facto information, it would be a worthwhile exercise to present the patterns (software components), whose utility class labels were adjusted, to the software architect in order to confirm that the class labels actually reflect the assignments as determined by the reference test.

8 Conclusion

The performance results from our experiments empirically demonstrate that centroid-adjusted class labelling is an effective preprocessing method for the utility analysis and classification of software components. By adjusting utility class labels, within a design subset, to reflect their proximity (similarity) to all class centroids, a concomitant performance gain was realized with the validation subset. Moreover, utility misclassifications of the software components tended to be more conservative. While adjustments may only be made to the design patterns, suspect validation patterns may be flagged by this method for subsequent analysis by a software architect. An avenue of future investigation would be to examine clustering strategies that would allow for several centroids for each utility class. Moreover, other quantile-based robust measures of dispersion may also be examined.

Acknowledgment The Natural Sciences and Engineering Research Council of Canada (NSERC) is gratefully acknowledged for its support of this investigation.

References

1. Adler, J.: R in a Nutshell, 2nd edn. O'Reilly Media Inc, Sebastopol (2012)
2. Aggarwa, C.C.: Data Classification: Algorithms and Applications. CRC Press, Boca Raton (2014)
3. Backus, J.: Can programming be liberated from the von Neumann style? A functional style and its algebra of programs. Commun. ACM **21**(8), 613–641 (1978)
4. Bishop, C.M.: Pattern recognition and machine learning. Springer, New York (2007)
5. Brown, B.M.: Statistical use of spatial median. J. Roy. Stat. Soc. B **45**, 25–35 (1983)
6. Canfora, G., Troiano, L.: The importance of dealing with uncertainty in the evaluation of software engineering methods and tools. In: Proceedings of the 14th International Conference on Software Engineering and Knowledge Engineering, Ischia, Italy, 15–19 July, pp. 691–698 (2002)
7. Card, D., Glass, R.: Measuring Software Design Quality. Prentice-Hall, Englewood Cliffs (1990)
8. Cesarini, F., Thompson, S.: Erlang Programming: A Concurrent Approach to Software Development. O'Reilly Media Inc, Sebastopol (2014)
9. Chidamber, S.R., Kemerer, C.F.: A metrics suite for object-oriented design. IEEE Trans. Softw. Eng. **20**, 476–493 (1994)

10. Church, A.: An unsolvable problem of elementary number theory. Am. J. Math. **58**, 345–363 (1936)
11. Coad, P., Mayfield, M., Kern, J.: Java Design: Building Better Apps & Applets. Prentice Hall, Upper Saddle River (1999)
12. Cover, T.M., Hart, P.E.: Nearest neighbor pattern classification. IEEE Trans. Inf. Theory **13**, 21–27 (1967)
13. Donoho, D.L.: Breakdown properties of multivariate location estimators. Ph.D. Qualifying Paper, Department of Statistics, Harvard University (1982)
14. Dougherty, G.: Pattern Recognition and Classification: An Introduction. Springer, New York (2013)
15. Duda, R.O., Hart, P.E., Stork, D.G.: Pattern Classification, 2nd edn. Wiley-Interscience, Hoboken (2004)
16. El-Alfy, E.-S.M., Thampi, S.M., Takagi, H., Piramuthu, S., Hanne, T.: Advances in Intelligent Informatics. Springer, Berlin (2014)
17. Emerick, C., Carper, B., Grand, C.: Clojure Programming: Practical Lisp for the Java World. O'Reilly Media Inc, Sebastopol (2012)
18. Everitt, B.S.: Moments of the statistics kappa and weighted kappa. Br. J. Math. Stat. Psychol. **21**(1), 97–103 (1968)
19. Fenton, N.E., Kaposi, A.A.: Metrics and software structure. Inf. Softw. Technol. **29**, 301–320 (1987)
20. Fenton, N.E., Pfleeger, S.L.: Software Metrics: A Rigorous and Practical Approach. PWS Publishing, Boston (1997)
21. Fleiss, J.L.: Measuring agreement between judges on the presence or absence of a trait. Biometrics **31**(3), 651–659 (1975)
22. Ford, N.: Functional Thinking: Paradigm Over Syntax. O'Reilly Media Inc, Sebastopol (2014)
23. Fowler, M.: Refactoring: Improving the Design of Existing Code. Addison-Wesley, Reading (1999)
24. Glover, F.: Tabu search, I. ORSA J. Comput. **1**, 190–206 (1989)
25. Grandvalet, Y., Canu, S.: Adaptive scaling for feature selection in SVMs. In: Advances in Neural Information Processing Systems, vol. 15 (NIPS 2002), pp. 569–576. Cambridge, MIT Press (2003)
26. Haldane, J.B.S.: Note on the median of a multivariate distribution. Biometrika **35**(3–4), 414–415 (1948)
27. Halstead, M.H.: Elements of Software Science. Elsevier, New York (1977)
28. Hastie, T., Tibshirani, R., Friedman, J.: The Elements of Statistical Learning: Data Mining, Inference, and Prediction, 2nd edn. Springer, New York (2011)
29. Henderson-Sellers, B.: Object-Oriented Metrics: Measures of Complexity. Prentice Hall, Upper Saddle River (1995)
30. Hoaglin, D.C., Mosteller, F., Tukey, J.W.: Understanding Robust and Exploratory Data Analysis. Wiley-Interscience, New York (2000)
31. Huang, S.-J., Lin, C.-Y., Chiu, N.-H.: Fuzzy decision tree approach for embedding risk assessment information into software cost estimation model. J. Inf. Sci. Eng. **22**, 297–313 (2006)
32. Huber, P.J.: Robust estimation of a location parameter. Ann. Math. Stat. **35**(1), 73–101 (1964)
33. Hudak, P., Jones, M.P.: Haskell vs. Ada vs. C++ vs. Awk vs. … An experiment in software prototyping productivity 1994, 17 p. http://haskell.cs.yale.edu/wp-content/uploads/2011/03/HaskellVsAda-NSWC.pdf
34. Hughes, J.: Why functional programming matters. Comput. J. **32**(2), 98–107 (1989)
35. Jones, C.: Software metrics: good. Bad Missing Comput. **27**, 98–100 (1994)
36. Jung, H.-W., Kim, S.-G., Chung, C.-S.: Measuring software product quality: a survey of ISO/IEC 9126. IEEE Softw. **21**, 88–92 (2004)
37. Kasabov, N., Song, Q.: DENFIS: dynamic evolving neural-fuzzy inference system and its application for time-series prediction. IEEE Trans. Fuzzy Syst. **10**, 144–154 (2002)

38. Kitchenham, B.A., Hughes, R.T., Kinkman, S.G.: Modeling software measurement data. IEEE Trans. Softw. Eng. **27**, 788–804 (2001)
39. Landis, J.R., Koch, G.G.: The measurements of observer agreement for categorical data. Biometrics **33**(1), 159–174 (1997)
40. Leroy, X., Doligez, D., Frisch, A., Garrigue, J., Rémy, D., Vouillon, J.: The OCaml system release 4.02: documentation and user's manual. Institut National de Recherche en Informatique et en Automatique (2014). http://caml.inria.fr/distrib/ocaml-4.02/ocaml-4.02-refman.pdf
41. Lieberherr, K.J., Holland, I.M.: Assuring good style for object-oriented programs. IEEE Softw. **6**, 38–48 (1989)
42. Liu, Q., Sung, A., Chen, Z., Xu, J.: Feature mining and pattern classification for LSB matching steganography in grayscale images. Pattern Recogn. **41**, 56–66 (2008)
43. Lyu, M.R.: Handbook of Software Reliability Engineering. McGraw-Hill, Toronto (1996)
44. Mangano, S.: Mathematica Cookbook. O'Reilly Media Inc, Sebastopol (2010)
45. Marinescu, R.: Detecting design flaws via metrics in object-oriented system. International Conference and Exhibition on Technology of Object-Oriented Languages and Systems, Santa Barbara, USA, 29 July–3 August, pp. 173–182 (2001)
46. McCabe, T.J.: A complexity metric. IEEE Trans. Softw. Eng. **2**, 308–320 (1976)
47. Mohri, M.: Foundations of Machine Learning. MIT Press, Cambridge (2012)
48. Murofushi, T., Sugeno, M.: A theory of fuzzy measures: Representations, the Choquet integral, and null sets. J. Math. Anal. Appl. **159**, 532–549 (1991)
49. O'Sullivan, B., Goerzen, J., Stewart, D.B.: Real World Haskell: Code You Can Believe In. O'Reilly Media Inc, Sebastopol (2008)
50. Okasaki, C.: Purely Functional Data Structures. Cambridge University Press, Cambridge (1998)
51. Pedrycz, W., Sosnowski, Z.A.: The design of decision trees in the framework of granular data and their application to software quality models. Fuzzy Sets Syst. **123**, 271–290 (2001)
52. Phelps, C.E., Hutson, A.: Estimating diagnostic test accuracy using a "fuzzy gold standard". Med. Decis. Mak. **15**(1), 44–57 (1995)
53. Pizzi, N.J.: Fuzzy preprocessing of gold standards as applied to biomedical spectra classification. Artif. Intell. Med. **16**(2), 171–182 (1999)
54. Pizzi, N.J.: Discrimination of biomedical patterns using centroid-adjusted class labels. Can. Appl. Math. Q. (2014, in press)
55. Poels, G., Dedene, G.: Distance-based software measurement: necessary and sufficient properties for software measures. Inf. Softw. Technol. **42**, 35–46 (2000)
56. Pressman, R.S., Maxim, B.R.: Software Engineering: A Practitioner's Approach, 8th edn. McGraw-Hill, New York (2014)
57. Pudil, P., Novovicová, J., Kittler, J.: Floating search methods in feature selection. Pattern Recogn. Lett. **15**, 1119–1125 (1994)
58. Reformat, M., Pedrycz, W., Pizzi, N.J.: Software quality analysis with the use of computational intelligence. Inf. Softw. Technol. **45**, 405–417 (2003)
59. Schmitt, E., Bombardier, V., Wendling, L.: Improving fuzzy rule classifier by extracting suitable features from capacities with respect to the Choquet integral. IEEE Trans. Syst. Man Cybern. **38**, 1195–1206 (2008)
60. Schölkopf, B., Smola, A.J.: Learning with Kernels: Support Vector Machines, Regularization, Optimization, and Beyond. MIT Press, Cambridge (2002)
61. Seber, G.A.F.: Multivariate Observations. Wiley, Hoboken (2007)
62. Sicilia, M.A., Cuadrado, J.J., Crespo, J., García-Barriocanal, E.: Software cost estimation with fuzzy inputs: fuzzy modeling and aggregation of cost drivers. Kybernetika **41**, 249–264 (2005)
63. Small, C.G.: Measures of centrality of multivariate and directional distributions. Can. J. Stat. **15**(1), 31–39 (1987)
64. Smith, C.: Programming F# 3.0: A Comprehensive Guide for Writing Simple Code to Solve Complex Problems, 2nd edn. O'Reilly Media, Inc., Sebastopol (2012)
65. Sommerville, I.: Software Engineering, 9th edn. Addison-Wesley, Boston (2010)

66. Sturm, O.: Functional Programming in C#: Classic Programming Techniques for Modern Projects. Wiley, Chichester (2011)
67. Tahir, M., Bouridane, A., Kurugollu, F.: Simultaneous feature selection and feature weighting using Hybrid Tabu Search/K-nearest neighbor classifier. Pattern Recogn. Lett. **28**, 438–446 (2007)
68. Tang, E.K., Suganthan, P.N., Yao, X.: Gene selection algorithms for microarray data based on least square support vector machine. BMC Bioinformatics **7**(95) (2006)
69. Theodoridis, S., Koutroumbas, K.: Pattern Recognition, 4th edn. Academic Press, San Diego (2008)
70. Tukey, J.W.: Mathematics and picturing data. In: Proceedings of the International Congress of Mathematicians, Vancouver, Canada, pp. 523–531 (1975)
71. Valenstein, P.N.: Evaluating diagnostic tests with imperfect standards. Am. J. Clin. Pathol. **93** (2), 252–258 (1990)
72. van den Berg, K.G., van den Broek, P.M.: Static analysis of functional programs. Inf. Softw. Technol. **37**(4), 213–224 (1995)
73. Vapnik, V.N.: The Nature of Statistical Learning Theory. Springer, New York (1995)
74. Vapnik, V.: Statistical Learning Theory. Wiley, New York (1998)
75. Vapnik, V., Lerner, A.: Pattern recognition using generalized portrait method. Autom. Remote Control **24**(6), 774–780 (1963)
76. Walter, S.D., Irwig, L.M.: Estimation of test error rates, disease prevalence, and relative risk from misclassified data: A review. J. Clin. Epidemiol. **41**(9), 923–937 (1988)
77. Wang, L.: Support Vector Machines: Theory and Applications. Springer, Berlin (2005)
78. Warburton, R.: Java 8 Lambdas: Functional Programming for the Masses. O'Reilly Media, Inc., Sebastopol (2014)
79. Weyuker, E.J.: Evaluating software complexity measures. IEEE Trans. Softw. Eng. **14**, 1357–1365 (1988)
80. Yehuda, V., Zhang, C.: The multivariate L1-median and associated data depth. Proc. Natl. Acad. Sci. **97**(4), 1423–1426 (2000)
81. Zadeh, L.A.: Outline of a new approach to the analysis of complex systems and decision processes. IEEE Trans. Syst. Man. Cybern. SMC-**3**(1), 28–44 (1973)

Author Biography

Nick Pizzi is the Chief Data Scientist with InfoMagnetics Technologies Corporation and Adjunct Professor with the University of Manitoba, Department of Computer Science. Dr. Pizzi has extensive experience with scientific computing, pattern recognition, machine learning, biomedical informatics, and multivariate statistical analysis. He analyzes and interprets complex, voluminous data in order to discover new generalizations, trends, and unanticipated patterns.

Toward Accurate Software Effort Prediction Using Multiple Classifier Systems

Bhekisipho Twala and June Verner

Abstract Averaging is a standard technique in applied machine learning for combining multiple classifiers to achieve greater accuracy. Such accuracy could be useful in software effort estimation which is an important part of software process management. To investigate the use of ensemble multiple classifiers learning in terms of predicting software effort. The use of ensemble multiple classier combination is demonstrated and evaluated against individual classifiers using 10 industrial datasets in terms of the smoothed error rate. Experimental results show that multiple classifier combination can improve software effort prediction with boosting, bagging and feature selection achieving higher accuracy rates. Accordingly, good performance is consistently derived from static parallel systems while dynamic classifier selection systems exhibit poor accuracy rates. Most of the base classifiers are highly competitive with each other. The success of each method appears to depend on the underlying characteristics of each of the ten industrial datasets.

Keywords Multiple classifiers · Machine learning · Software effort · Predictive accuracy

B. Twala (✉)
Department of Electrical and Electronic Engineering Science, University of Johannesburg,
P.O. Box 524, Auckland Park, Johannesburg 2006, South Africa
e-mail: btwala@uj.ac.za

J. Verner
Computer Science and Engineering, University of New South Wales UNSW,
Sydney, NSW 2052, Australia
e-mail: june.verner@gmail.com

© Springer International Publishing Switzerland 2016   135
W. Pedrycz et al. (eds.), *Computational Intelligence and Quantitative
Software Engineering*, Studies in Computational Intelligence 617,
DOI 10.1007/978-3-319-25964-2_7

1   Introduction

Software effort estimation is an important area for software development. If the software development effort is under estimated tight time schedules will result leading to the possibility of inadequate testing and poor quality software. In contrast, if the software development effort is overestimated over allocation of man power and resources may result. Thus, accurate software effort estimation is an important part of the software management process in terms of productivity and quality. Many software effort estimation models have been proposed [2, 6, 18] and unbiased effort prediction is an important contributor to effective software project management. It is also generally accepted that the highest accuracy results that a classifier system can achieve depend on the quality of data and the appropriate selection of a learning algorithm for the data [7, 27, 35]. One of the central tasks of classifiers is determining whether a particular instance belongs to a specified class, given a description of that instance. The wealth and complexity of industrial data lends itself well to the application of classifiers for prediction or classification of software projects according to factors that influence software effort rates.

Machine learning (ML) deals with the problem of building computer programs that improve their performance at some tasks through experience and has proven to be of great value in a variety of applications including software development effort estimation (the process of predicting the effort required to develop a software system). In recent years, several machine learning approaches have been applied in software systems development and deployment in order to establish more sound predictive models for software quality [41]. Averaging is a standard technique in applied and theoretical ML for combining multiple classifiers in order to achieve great accuracy. In fact, in recent years, there has been an explosion of papers in the ML and statistical pattern recognition (SPR) communities discussing how to combine models or model predictions in order to improve predictive accuracy.

Research in both ML and SPR communities has shown that combining (ensemble) individual classifiers is an effective technique for improving predictive accuracy. In other words, developing an effective decision combination function is critical to the success of a multiple classifier system (MCS). Such a function should take advantage of the strengths of individual classifiers while avoiding their weaknesses, and improve classification correctness. The performance of multiple classifier systems not only depends on the power of the individual classifiers in the system but is also influenced by the independence between classifiers.

It has long been recognized that software effort estimation is a key consideration for good software cost estimation. However, effort prediction in terms of using multiple classifier (machine) learning or ensembles has attracted some attention in areas such as pattern recognition [24, 25], information security [8], credit risk [37], engineering [36], and so on, but yet received little attention in the software engineering community. Work by Wettschereck [26] provides a solid start to the use of multiple classifier learning by proposing a hybrid strategy that combines the nearest-hyper-rectangle and *k*-nearest neighbour algorithms in terms of improving classification accuracy.

Follow up research work by Braga et al. [3] and Kultur et al. [23] shows how bagging may improve software effort predictive accuracy in comparison with the use of a single classifier; although the results from both studies are inconclusive. Khoshgoftaar et al. [19] propose a hybrid software quality prediction model that combines rule-based and case-based learning which outperforms the best individual rule-based model. Kocaguneli et al. [21] suggest that ensembles are not able to improve predictive accuracy of single learning classifiers, contradicting the findings of Khoshgoftaar et al.'s research. However, the Kocaguneli et al.'s [21] research work lacks any statistical justification. In their most recent research work, Kocaguneli et al. [22] show ensemble methods significantly outperforming single classifiers with error rates significantly less than are shown by their earlier work. The ranking of the best ensemble methods were also shown to be stable by Kocaguneli et al. [22]. Twala and Cartwright [37] showed that the ensemble approach can also be used to improve software effort predictive accuracy in the presence of missing values.

The performance of several multiple classifier systems are evaluated in terms of their ability to predict software effort using 10 industrial datasets in this research. Initially single classifiers are constructed using five base methods for classifier construction. These are then used to provide benchmarks against which various multiple classifier systems are assessed. To the best of our knowledge this is the first study where such a combination of methods in terms of classifier learning and ensemble learning approaches have been used to create different ensemble multiple classifier systems across ten industrial datasets. A classifier ensemble is generated by training multiple learners for the same task and then combining their predictions as demonstrated in Sect. 3 of the paper. There are different ways in which ensembles can be generated, and the resulting output combined for the classification of new instances. Popular approaches for creating ensembles include changing the instances used for training through techniques such as bagging [4], boosting [13], stacked generalization or stacking [40], changing the features used in training [15], and introducing randomness in the classifier itself [10].

Bagging is a combination of bootstrapping and averaging used to decrease the variance part of prediction errors; boosting is one of the most well-known techniques for solving classification problems; stacking combines various machine learning methods using a stacking generalization technique; randomization is based on bagging models built using a random tree strategy in which classification trees are grown on a random subset of descriptors; feature selection aims for an optimal set as a whole rather than a combination of stand-alone high performance attributes.

The rest of this paper is organised as follows. Section 2 briefly provides details of the five classifiers used in this paper; this is followed by a description of different types of multiple classifier system architectures. Section 4 empirically explores the robustness and accuracy of five multiple classifier systems when used with ten industrial datasets in terms of the smoothed error rate. This section also presents empirical results from the application of the ensemble procedures. Section 5 provides our conclusions and future research directions.

2   Classifiers

In supervised learning, for multivariate data, a classification function $y = f(x)$ from training examples of the form $\{(x_1, y_1), \ldots, (x_m, y_m)\}$, predicts one (or more) output attribute(s) or dependent variable(s) given the values of the input attributes of the form $(x, f(x))$. The x_i values are vectors of the form $\{x_{i1}, \ldots, x_{in}\}$ whose components can be numerically ordered, nominal or categorical, or ordinal. The y values are drawn from a discrete set of classes $\{1, \ldots, K\}$ in the case of *classification*. Depending on the usage, the prediction can be "definite" or probabilistic over possible values. Given a set of training examples and any given prior probabilities and misclassification costs, a learning algorithm outputs a *classifier*. The classifier is an hypothesis about the true classification function that is learned from, or fitted to, *training data*. The classifier is then tested on *test data*.

The five base methods for classifier construction considered in our study are presented below.

2.1   *Logistic Discrimination*

Logistic discrimination analysis (LgDA) Cox [9] is related to logistic regression. The dependent variable can only take the values 0 and 1, say, given two classes. This technique is partially parametric, as the probability density functions for the classes are not modelled but rather the ratios between them are used as described below.

Let $y \in \{0, 1\}$ be the dependent or response variable and let $x = x_{i1}, \ldots, x_{ip}$ be the predictor variables vector. A linear predictor ζ_i is given by $\beta_0 + \beta'_x$ where β_0 is the constant and β' is the vector of regression coefficients $(\beta_1, \ldots, \beta_p))$ to be estimated from the data. They are directly interpretable as log-odds ratios or in terms of $exp(\beta')$, as odds ratios.

The a posteriori class probabilities are computed by the logistic distribution. These terms are often referred to as "predictions" for the given characteristic vector x. Therefore, a new element is classified as 0 if $\pi_0 \leq c$ and as 1 if $\pi_0 > c$, where c is the cut-off point score and π_0 is the predictor. Typically, the error rate is lowest for cut-off point = 0.5 [30]. In fact, the slope of the cumulative logistic probability function has been shown to be steepest in the region where, say, $\pi_i = 0.5$. Thus, if $\pi_i > 0.5$, the unknown instance is classified as "1" and if $\pi_i \leq 0.5$, the unknown instance is classified as "0". The generalisation of the LgDA approach to the case of three or more classes is known as the multinomial logit model and the derivation is similar to that of the logistic discrimination model. The reader is referred to Hosmer and Lameshow [16] for more details.

2.2 k-Nearest Neighbour

One of the most accepted algorithms in ML is the k-nearest neighbour (k-NN), which is sometimes referred to as instance-based learning or memory-based reasoning [1]. k-NN methods have been used for classification tasks. The method essentially works by assigning to an unclassified sample point the classification of the nearest of a set of previously classified points. The entire training set (a set of data used to discover potentially predictive relationships in different areas of information science) is stored in the memory. Consider a set of n pairs is $(x_1, C_1), \ldots, (x_n, C_n)$, where x_i's take values in the metric space X upon which is defined a metric d, and the C_i's take values in the set $\{1, 2, \ldots, K\}$. A new measurement x is observed, and it is desired to estimate C by utilising the information contained in the set of correctly classified points. $x'_n \in \{x_1, \ldots, x_n\}$ is called a nearest neighbour to x if $\min d(x_i, x) = d(x'_n, x)$ i $= 1, 2, \ldots,$ n. The nearest neighbour rule decides that x belongs to the category C'_n of its nearest neighbour x'_n. A mistake is made if $C'_n \neq C$. Notice that only classification of the nearest neighbour is utilised by this, simplest, nearest neighbours rule. The remaining $n - 1$ classifications C_i are ignored.

To classify a new instance, the Euclidean distance (possibly weighted) is computed between the instance and each stored training instance and the new instance is assigned the class of the nearest neighbouring instance. More generally, these k-nearest neighbours (k-NNs) are computed, and the new instance is assigned the class that is most frequent amongst the k neighbours. IBL's have three defining general characteristics: a similarity function (how close together the two instances are), a "typical instance" selection function (which instances to keep as examples), and a classification function (deciding how a new case relates to the learned cases). The lack of a formal framework for choosing the size of neighbourhood "k" can be problematic. To determine the distance between a pair of instances we apply the Euclidean distance metric. In our experiments, k is set to five. Three to five neighbours have been shown to make a good prediction [38].

2.3 Artificial Neural Network

Artificial neural networks (ANNs) use nonparametric approaches (i.e. no assumptions about the data are made). ANNs are represented by connections between a very large number of simple computing processors or elements (neurons). ANNs have been used for a variety of classification and regression problems. There are many types of ANNs, but for the purposes of this study we concentrate on single unit and multi-layer perceptrons [29] which utilizes a supervised learning technique known as backpropagation.

The backpropagation learning algorithm performs a hill-climbing search procedure on the weight space described above or a (noisy or stochastic) gradient descent numerical method whereby an error function is minimised. At each iteration, each weight is adjusted proportionally to its effect on the error. One cycles through the training set and on each example changes each weight proportionally to its effect on lowering the error. One may compute the error gradient using the chain rule and the information propagates backwards through the network through the interconnections, which accounts for the procedure's name.

There are two stages associated with the backpropagation method: training and classification. The ANN is trained by supplying it with a large number of learned (input data pattern) whose corresponding classifications (target values or desired output) are known. During training, the final sum-of-squares error over the validation data for the network is calculated. The selection of the optimum number of hidden nodes is made on the basis of this error value. The question of how to choose the structure of the network is beyond the scope of this thesis and is a current research issue in neural networks. Once the network is trained, a new object is classified by sending its attribute values to the input nodes of the network, applying the weights to those values, and computing the values of the output units or output unit activations. The assigned class is that with the largest output unit activation.

2.4 Decision Trees

Decision tree (DT) classifiers have four major objectives. According to Safavian and Landgrebe [31], these are: (1) to classify correctly as much of the training sample as possible; (2) generalise beyond the training sample so that unseen samples could be classified with as high accuracy as possible; (3) be easy to update as more training samples become available (i.e., be incremental); (4) and have as simple a structure as possible. Objective (1) is actually highly debatable as this might not be the case and to some extent conflicts with objective (2). Also, not all tree classifiers are concerned with objective (3). DTs are non-parametric and a useful means of representing the logic embodied in software routines. A DT [5, 28] takes as input a case or example described by a set of attribute values, and outputs a Boolean or multi-valued "decision". For the purpose of this paper, we shall stick to the Boolean case.

One property that sets DTs apart from all other classifiers is their invariance to monotone transformations of the predictor variables. For example, replacing any subset of the predictor variables $\{x_j\}$ by (possible different) arbitrary strictly monotone functions of them $\{x_j \leftarrow m_j(x_j)\}$, gives rise to the same tree model. Thus, there is no issue with having to experiment with different possible transformations $m_j(x_j)$ for each individual predictor x_j to try to find the best. This invariance provides immunity to the presence of extreme values ("outliers" or noise) in the predictor variable space [5].

2.5 Naïve Bayes Classifer

The NBC is perhaps the simplest and most widely studied probabilistic learning method. It learns from the training data the conditional probability of each attribute A_i given the class label C [11]. The NBC can handle an arbitrary number of independent attributes whether continuous or categorical. The strong major assumption is that all attributes A_i are independent given the value of the class C. Classification is therefore done applying Bayes rule to compute the probability of, say, C given $A_1,\ldots,A_n$ and then predicting the class with the highest posterior probability. The probability of a class value C_i given an instance $X = \{A_1,\ldots,A_n\}$ for n observations is given by:

$$p(C_i|X) = p(X|C_i) \cdot p(C_i)/p(X)$$
$$\propto p(A_1,\ldots,A_n|C_i) \cdot p(C_i)$$
$$= \prod_{j=1}^{n} p(A_j|C_i) \cdot p(C_i)$$

The assumption of conditional independence of a collection of random variables is very important for the above result. It would be impossible to estimate all the parameters without such an assumption. This is a fairly strong assumption that is often not applicable. However, bias in estimating probabilities may not make a difference in practice—it is the order of the probabilities, not the exact values that determine the probabilities. When the strong attribute independence assumption is violated, the performance of the NBC might be poor.

3 Multiple Classifier System Architectures

Multiple classifier systems can be classified into one of three architectural types [12]: (1) static parallel (SP); (2) multi-stage (MS); and (3) dynamic classifier selection (DCS). The outputs from each classifier are combined to deliver a final classification decision. A large number of combination functions are available. These include: voting methods (simple majority vote, weighted majority vote, the product or sum of model outputs also known as the product rule, the minimum rule, the maximum rule); rank based methods (borda-count); and probabilistic methods (Bayesian methods).

3.1 Static Parallel

SP is probably the most popular architecture and it is where two or more classifiers are developed independently in parallel [42]. The outputs from each classifier are then combined to deliver a final classification decision (where the decision is selected from a set of possible class labels). A large number of combination functions are available. These include majority voting, weighted majority voting, the product or sum of model outputs, the minimum rule, the maximum rule and Bayesian methods. In practice most combination strategies are reported to yield very similar levels of performance. However, a simple majority vote or weighted majority vote are often favoured due to the simplicity of their application and their applicability to situations where the raw outputs from each classifier may not all be interpretable in the same way.

3.2 Multi-stage

The second type of architectures is MS, where the classifiers are constructed iteratively. At each iteration (and at previous stages), the parameter estimation process is dependent upon the classification properties of the classifier(s) developed. Some MS approaches generate models that are applied in parallel using the same type of combination rules used for SP methods. For example, most forms of boosting generate a set of weak classifiers that are combined to create stronger ones [33]. Adaboost [13] is one of the most well-known algorithms that uses a MS architecture.

3.3 Dynamic Classifer Selection

For DCS, different classifiers are developed or applied to different regions within the problem domain. While one classifier may be shown to outperform all others based on global measures of performance, it may not entirely dominate all other classifiers. Weaker competitors will sometimes outperform the overall best across some regions [20]. DCS problems are normally approached from a global and local accuracy perspective [24, 25]. With a DCS global approach classifiers are constructed using all observations within the development sample. Classifier performance is then assessed over each region on interest (I am not sure what this term means) and the best classifier is chosen for each region. With DCS local, regions of interest are determined first, and then separate classifiers are developed for each region.

3.4 Classifier Ensemble

A generalised classifier ensemble algorithm is summarised in the following steps [34].

1. Partition original dataset into n training datasets, TR_1, TR_2, TR_n.
2. Construct n individual models (M_1, M_2, M_n) with the different training datasets TR_1, TR_2, ..., TR_n to obtain n individual classifiers (ensemble members) generated by different algorithms, thus different.
3. Select m de-correlated classifiers from n classifiers using de-correlation maximization algorithm.
4. Using Step 3, obtain m classifier output values (misclassification error rates) of unknown instance.
5. Transform output value to reliability degrees of positive class and negative class, given the imbalance of some datasets.
6. Fuse the multiple classifiers into aggregate output in terms of majority voting.

4 Experimental Design

In order to test the suitability of multiple classifiers for predicting software effort, w performed experiments on ten industrial datasets in terms of the smoothed mis-classification error rate. The smoothed error rate is used due to its variance reduction benefit. Instead of summing terms that are either zero or one as in the error-count estimator, the smoothed estimator uses a continuum of values between zero and one in the terms that are summed. The resulting estimator has a smaller variance than the error-count estimate. Each dataset, used in the experiments defines a different learning problem as summarized in Table 1. Most of the datasets are available at predictor models in software engineerinig (PROMISE) [32] with the exception of ISBSG and Company X which is not available for public use due to non-disclosure agreement.

For the simulation study, the five base methods of classifier construction were chosen. Each method utilizes a different form of parametric estimation/learning; between them they generate different models forms: linear models, density estimation, trees and networks; and they are all practically applicable within software engineering environments, with known examples of their application within the engineering management industry. To begin, single classifiers were constructed using each method. These were used to provide benchmarks against which various

Table 1 Industrial datasets problem

Dataset	Instances	Attributes		Mean development effort
		Numerical	Categorical	
Test equipment	16	17	4	236
Kemerer	18	4	2	261
Test equipment	16	17	4	379
Bank	18	2	7	1470
Test equipment	16	17	4	550
Data science institute	26	5	0	2528
Moser	32	1	1	2874
Desharnais	77	3	6	4834
Experience	95	1	5	1443
ISBSG-version 7	166	2	7	1668
China	499	16	2	3921
Company X	10,434	4	18	41,643

multiple classifier systems were assessed. To select an appropriate number of ensemble members, the de-correlation maximization method [17] was utilized. 10-fold cross validation is used for all the experiments.

For all the classifiers, the implementation in WEKA data mining software package library [39] is used, with the default parameters used for each classifier. These models were built in WEKA by performing five-fold cross validation.

Analyses of variance are used to examine the main effect and their respective interactions. This was done using a 3-way repeated measures design (where each effect was tested against its interaction with datasets). The fixed effect factors are multiple classifier methods; the ensemble learning approaches used to build the multiple classifier systems and the multiple classifier architectures. The random effect is the ten datasets. Friedman ranking test [14] was also used to check if the difference in performances between the multiple classifiers (ensembles) and the individual classifiers were significantly different in terms of the smoothed error rate.

To measure the performance of classifiers, the training set/test set methodology is employed. For each run, each dataset is split randomly into 80 % training set and 20 % testing or validation set. The performance of each classifier is then assessed on the smoothed error rate.

Although, an operational definition of accurate prediction is hard to come by predictive accuracy is mostly operationally defined as the prediction with the minimum misclassification costs (the proportion of misclassified instances). The need for minimizing costs, rather than the proportion of misclassified instances, arises when some predictions that fail are more catastrophic than others, or when some predictions that fail occur more frequently than others. Minimizing costs, however, does correspond to minimizing the proportion of misclassified instances when priors (i.e. the probability estimates drawn from the training data that one would make for each possible target value prior to knowing anything about the predictor values) are taken to be proportional to the class sizes and when mis-classification costs are taken to be equal for every class [5]. This is the approach we follow in the paper.

5 Experimental Results

The results across all the ten datasets are summarized in Figs. 1, 2, 3 and 4 (and Tables 2, 3 and 4) in terms of smoothed error rate against the baseline classifiers (BASE) and their respective ensemble multiple classifiers (i.e. ENS1, ENS2, ENS3, ENS4, ENS5). The components of the ensembles are ENS1 (ANN, DT, NBC,

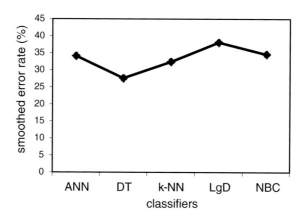

Fig. 1 Overall means for base classifiers

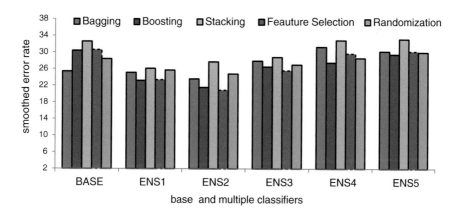

Fig. 2 Ensemble multiple classifiers (static parallel)

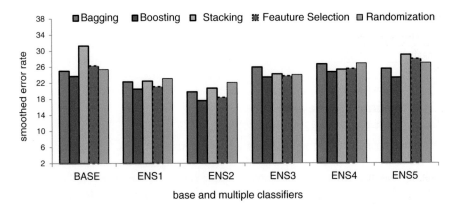

Fig. 3 Ensemble multiple classifiers (multi-stage)

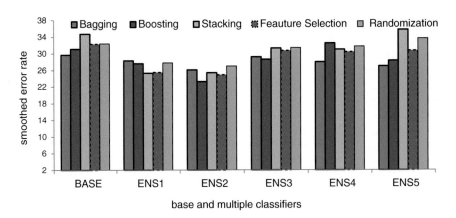

Fig. 4 Ensemble multiple classifiers (dynamic classifier selection)

Table 2 Overall means (individual classifiers and multiple clasifier systems)

Classifier/ensemble multiple classifiers	Average generalization performance (%)
ANN	34.1 ± **3.63**
DT	27.6 ± **3.99**
NBC	35.6 ± **3.27**
k-NN	32.4 ± **3.51**
LgD	38.1 ± **3.90**
ENS1	20.4 ± **2.95**
ENS2	17.9 ± **1.75**
ENS3	22.0 ± **3.23**
ENS4	23.7 ± **3.27**
ENS5	25.6 ± **3.61**

Table 3 Overall means (ensemble learning approaches)

Learning approaches	Average generalization performance (%)
Bagging	16.5 ± 2.82
Boosting	19.4 ± 3.32
Feature selection	20.3 ± 3.51
Stacking	26.9 ± 4.73
Randomization	22.1 ± 3.38

Table 4 Overall means (multiple classifier architectures)

Multiple classifier architectures	Average generalization performance (%)
Dynamic classifier selection	24.2 ± 3.64
Multi stage	21.9 ± 3.41
Static parallel	17.4 ± 2.95

k-NN, LgD) ENS2 (ANN, DT, NBC, LgD) ENS 3(ANN, DT, k-NN, LgD) ENS4 (ANN, NBC, k-NN, LgD) and ENS5 (DT, NBC, k-NN, LgD).

For the baseline classifiers, DT achieves the lowest smoothed error rate (27.6 %), followed by k-NN (32.4 %), ANN (34.1 %) and NBC (35.6 %), respectively. The worst performance for predicting software effort is by LgD with a smoothed error rate of 38.1 %. The differences in performance between the base individual classifiers are significant at the 5 % level (with the exception of ANN against NBC).

From Table 2 the multiple classifier performances (ENS1, ENS2, ENS3, ENS4, ENS5) is significantly better when compared to the individual classifiers (ANN, DT, NBC, k-NN, LgD) at the 95 % level of significance ($F_{statistic} = 13.141 > F_{critical\ value(10,\ 75)} = 2.056$).

When comparing the smoothed error rates in Table 3, bagging outperforms all the other sampling methods whenever it is used to construct the ensemble multiple classifier systems in software effort prediction. Bagging exhibits a smoothed error rate of 16.5 %, followed by boosting (19.4 %) and feature selection (20.3 %). However, there appears to be no significant difference in performance between boosting and feature selection at the 5 % level. Poor performance is observed when stacking is used, achieving a smoothed error rate of 26.9 %.

From Table 4, the results show that static parallel ensemble multiple classifier systems performs better in terms of predicting software effort when compared with either dynamic classifier selection or multi-stage systems. The difference in performance between the three systems is significantly different at the 5 % level.

All the static parallel systems (Fig. 2) show some potential to significantly outperform the baseline. However, stacking and bagging are the weakest, with only ensembles using ANN, LgD and DT showing major improvement over the other multiple classifier architectures.

Multi-stage systems provide statistically significant benefits over baseline models. The clear winners are feature selection and boosting, which provide large and significant improvements over the baseline and other multiple classifier systems for all methods considered, with best performance when applied to NBC (Fig. 3).

DCSs that look to segment the population into a number of sub-regions are consistently poor performers, with all the experiments yielding results that are inferior to the single best classifier. However, the performance of most static parallel and multi-stage combination strategies provide statistically significant improvements compared to DCSs (Fig. 4).

6 Conclusion

Machine learning has proved to be promising for automating software development effort. The rise of big data is most likely the largest catalyst. There are other factors as well that have made machine learning algorithms faster and easier to run which has been of great benefit to engineers. Not only does it enable the replication of results it provides some of the much needed automation capability in terms of engineering analysis and automated process planning. This covers design of software development processes in a wide range of domains. Multiple classifier learning could be involved here in areas such as learning process plans, learning error recovery strategies, learning and models for physical processes, and so on.

In this chapter we have proposed a strategy that uses machine learning techniques to improve software effort predictive accuracy. In summary, it has been found that a combination of multiple classifiers can enhance the classification and prediction accuracy of software effort to a great extent. Based on the experiments and findings on this paper, it can be concluded that multiple classifier combination can play an important role in the accurate and unbiased prediction of software effort by making full use of the abundant and detailed information in software projects and integrating the benefits of different classifiers. Thus, we can conclude that practitioners and researchers may use bagging and boosting for constructing models to predict software effort especially when measuring the quality of systems in software development. In fact, multiple classifier learning provides a new style of software development. But there are still many issues for further study, for example, developing models that would identify trends in effort revisions, selection of larger datasets, selection of member classifier, optimization of feature sets and determination of combination strategy. We intend to present our future findings in the next research journal paper.

Acknowledgments The project was sponsored by the Department of Electrical and Electronic Engineering Science at the University Of Johannesburg, South Africa. The authors would like to thank their colleagues for their valuable comments and suggestions to improve the paper.

References

1. Aha, D.W., Kibbler, D., Albert, M.K.: Instance-based learning algorithms. Mach. Learn. **6** (37), 37–66 (1991)
2. Basha, S., Dhavechelvan, P.: Analyisis of empirical software effort estimation models. Int. J. Comput. Sci. Inf. Secur. **7**(3), 68–77 (2010)
3. Braga, P.L., Oliveira, A., Ribeiro, G., Meira, S.: Bagging predictors for estimation of software project effort. In: International Joint Conference on Neural networks, Orlando, pp. 1595–1600 (2007)
4. Breiman, L.: Bagging predictors. Mach. Learn. **26**(2), 123–140 (1996)
5. Breiman, L., Friedman, J., Olshen, R., Stone, C.: Classification and regression trees. Wadsworth (1984)
6. Briand, L.C., Wieczorek, I.: Resource estimation in software engineering. In: Marcinak, J. J. (ed.) Encyclopedia of Software Engineering, pp. 1160–1196. Wiley, New York (2002)
7. Brodley, C.E., Friedl, M.A.: Identifying mislabeled training data. J Artif. Intell. Res. **11**, 131–167 (1999)
8. Corona, I., Giacinto, G., Roli, F.: Intrusion detection in computer systems using multiple classifier systems. In: Okun, O., Valentini, G. (eds.) Supervised and Unsupervised Ensemble Methods and Their Applications, vol 126, pp. 91–114. Springer, Berlin (2008)
9. Cox, D.R.: Some procedures associated with the logistic qualitative response curve. In: David, F.N. (ed.) Research Papers in Statistics: Festschrift for J. Neyman, pp. 55–71. Wiley, New York (1966)
10. Dietterich, T.: An experimental comparison of three methods for constructing ensembles of decision trees: bagging, boosting, and randomization. Mach. Learn. **40**(2), 139–158 (2000)
11. Duda, R.O., Hart, P.E.: Pattern Classification, 2nd edn. Wiley, New York (1973)
12. Finlay, S.M.: Multiple classifier architectures and their application to credit risk assessment. Working Paper 2008/012, Department of Management Science, Lancaster University, UK (2008)
13. Freund, Y., Schapire, R.: A decision theoretic generalization of on-line learning and an application to boosting. J. Comput. Syst. **55**, 119–139 (1996)
14. Friedman, M.: The use of ranks to avoid the assumption of normality implicit in the analysis of variance. J. Am. Stat. Assoc **32**(200), 675–701 (1937)
15. Ho, T.K.: Random decision forests. In: Proceedings of the 3rd International Conference on Document Analysis and Recognition, pp. 278–282 (1995)
16. Hosmer, D.W., Lameshow, S.: Applied Logistic Regression. Wiley, New York (1989)
17. Jolliffe, I.: Principal Component Analysis. Springer, Berlin (1986)
18. Jørgensen, M.: A review of studies on expert estimation of software development effort. J. Syst. Softw. **70**(1–2), 37–60 (2004)
19. Khoshgoftaar, T.M., Xiao, Y., Gao, K.: Software quality assessment using a multi-strategy classifier. Inf Sci (2010, in press)
20. Kittler, J., Hatef, M., Duin, R.P.W., Matas, J.: On combining classifiers. IEEE Trans. Pattern Anal. Mach. Intell. **20**(3), 226–239 (1998)
21. Kocaguneli, E., Bener, A., Kultur, Y.: Combining multiple learners induced on multiple datasets on software effort prediction. In: International Symposium on Software Reliability Engineering, Mysuri, India, p. 6 (2009)
22. Kocaguneli, E., Menzies, T., Keung, J.: On value of ensemble effort estimation. IEEE Trans. Softw. Eng. **38**(06), 1403–1416 (2012)
23. Kultur, Y., Turhan, B., Bener, A.: Ensemble of neural networks with associative memery (ENNA) for estimating software development costs. Knowl. Based Syst. **22**, 395–402 (2009)
24. Kuncheva, L.I.: Swithcing between selection and fusion in combining classifers: an experiment. IEEE Trans. Syst. Man Cybern. Part B Cybern. **32**(2), 146–156 (2002)
25. Kuncheva, L.: A theoretical study in six classifier fusion strategies. IEEE Trans. Pattern Anal. Mach. Intell. **24**(2), 281–286 (2002)

27. Pechenizkiy, M., Tsymbal, A., Puuronen, S., Pechenizkiy, O.: Class noise and supervised learning in medical domains: the effect of feature extraction. In: Proceedings of the 19th IEEE Symposium on Computer-Based Medical Systems, pp. 708–713 (2006)
28. Quinlan, J.R.: C4.5: Programs for Machine Learning. Los Altos, California. Morgan Kauffman Publishers INC, Burlington (1993)
29. Ripley, B.D.: Pattern Recognition and Neural Networks. Cambridge University Press, Cambridge and Wiley, New York (1992)
30. Rumelhart, D.E., Hinton, G.E., Williams, R.J.: Learning internal representations by error propagation. In: Rumelhart, D.E., McClelland, J.L. (eds.) Parallel Distributed Processing, vol. 1, pp. 318–362. MIT Press, Cambridge (1986)
31. Safavian, S.R., Landgrebe, D.: A survey of decision tree classifier methodology. IEEE Trans. Syst. Man Cybern. **21**, 660–674 (1991)
32. Sayyad, J.S., Menzies, T.J.: The PROMISE repository of software engineering databases. School of Information Technology and Engineering, University of Ottawa, Canada (2005). http://promise.site.uottawa.ca/SERepository. Accessed on 01 Dec 2014
33. Schapire, R., Freund, Y., Bartlett, P., Lee, W.: Boosting the margin: a new explanation for the effectiveness of voting methods. In: Proceedings of International Conference on Machine Learning, Morgan Kaufmann, San Francisco pp. 322–330 (1997)
34. Twala, B.: multiple classifier application to credit risk assessment. Expert Syst. Appl. **37**(4), 3236–3336 (2010)
35. Twala, B.: Effective techniques for dealing with incomplete data using decision trees. Published PhD thesis, Open University, Milton Keynes, UK (2005)
36. Twala, B.: software faults prediction using multiple classifiers. In: IEEE International Conference on Computer Research and Development (ICCRD2011), Shanghai, China, 11–13 Mar 2011
37. Twala, B., Cartwright, M.: Ensemble missing data methods in software effort prediction. Intell. Data Anal. **14**, 299–331 (2010)
38. Venables, W., Ripley, B.: Modern Applied Statistics with S-Plus. Springer, Berlin (1997)
26. Wettschereck, D.: A hybrid nearest neighbour and nearest hyperrectangle algorithm. In: Bergadano, F., Raedt, L.D. (eds.) Proceedings of European Conference on Machine Learning, pp 323–335 (1994)
39. Witten, I., Frank, E.: Data Mining Practical Machine Learning Tools and Techniques, 2nd edn. Morgan Kauffman, Burlington (2005)
40. Wolpert, D.: Stacked generalization. Neural Netw. **5**(2), 241–259 (1992)
41. Zhang, D., Tsai, J.J.P.: Advances in Machine Learning Applications in Software Engineering (2007)
42. Zhu, H., Beling, P.A., Overstreet, G.A.: A study in the combination of two consumer credit scores. J. Oper. Res. Soc. **52**, 2543–2559 (2001)

Author Biography

Bhekisipho Twala is a Professor in Artificial Intelligence and Statistical Sciences and the Head of the Electrical and Electronic Engineering Science Department at the University of Johannesburg. Before then he was a Principal Research Scientist at the Council of Science and Industrial Research (CSIR) within the Modelling and Digital Science unit. His current work involves promoting and conducting research in artificial intelligence within the electrical and electronic engineering fields and developing novel and innovative solutions to key research problems in this field. He earned his B.A. in Economics and Statistics from the University of Swaziland in 1993; his MSc in Computational Statistics from Southampton University (UK) in 1995; and his Ph.D. in Machine Learning and Statistics from the Open University (UK) in 2005. Prof. Twala was a post-doctoral researcher and Bournemouth University (UK) and later at Brunel University in the UK, mainly focussing on empirical software engineering research. His broad research interests include multivariate statistics, classification methods, knowledge discovery and reasoning with uncertainty, sensor data fusion and inference, and the interface between statistics and computing. He has particular interests in applications in finance, medicine, psychology, software engineering and most recently in robotics and has published over 70 scientific papers. Prof. Twala has a wide ranging work experience to organisations ranging from banks, through universities, to governments. He is currently an associate editor of the Intelligent Data Analysis journal, Journal of Computers, International Journal of Advanced Information Science and Technology, International Journal of Big Data Intelligence, Journal of Image and Data Fusion, Journal of Information Processing Systems, and a fellow of the Royal Statistical Society. Other professional memberships include the Association of Computing Machinery (ACM); the Chartered Institute of Transport (CIT), South Africa and a senior member of the Institute of Electrical and Electronics Engineers (IEEE).

Complex Fuzzy Logic Reasoning-Based Methodologies for Quantitative Software Requirements Specifications

Dan E. Tamir, Carl J. Mueller and Abraham Kandel

Abstract Quantitative software engineering is one of the most important paradigms for software development. That is, Requirements, Analysis, Design, Coding, and Testing. One of the challenges associated with quantitative software engineering is the fact that many of the quantifiable parameters are concomitant with uncertainty. Part of the uncertainty is due to the fact that a significant portion of the software engineering process involves human beings presenting rational, yet difficult to quantify, behavior. Due to this fact, soft computing approaches, specifically fuzzy logic based reasoning, present significant opportunities for constructing sound quantitative software engineering models. This work presents a new and innovative approach for fuzzy logic based quantitative software engineering procedures. We present a complex fuzzy logic based inference system used to account for the intricate relations between software engineering constraints such as quality, software features, and development effort. The new model concentrates on the requirements specifications part of the software engineering process. Moreover, the new model significantly improves the expressive power and inference capability of the soft computing component in the soft computing based quantitative software engineering.

Keywords Fuzzy logic · Fuzzy set theory · Fuzzy inference · Complex fuzzy logic · Complex fuzzy set theory · Complex fuzzy inference · Software engineering · Software development · Software requirements specifications · Quantitative software engineering · Quantitative software development · Quantitative software requirements specifications · Computational intelligence

D.E. Tamir (✉)
Department of Computer Science, Texas State University, San Marcos, TX, USA
e-mail: dt19@txstate.edu

C.J. Mueller
Department of Computer Information Systems, Texas A&M University—Central Texas,
Killeen, TX, USA

A. Kandel
School of Computing and Information Sciences, Florida International University,
Miami, FL, USA

© Springer International Publishing Switzerland 2016
W. Pedrycz et al. (eds.), *Computational Intelligence and Quantitative Software Engineering*, Studies in Computational Intelligence 617,
DOI 10.1007/978-3-319-25964-2_8

153

1   Introduction

Since Software Engineering's introduction in 1968, one of the challenges facing practitioners is to eliminate the uncertainties arising from the chaotic nature of software development [1]. One of the most widely known outcomes of the first International Conference on Software Engineering in Garmisch, Germany was the detection of a gap between the available design and implementation practices and the complexity of the software under development [1]. The gap and crisis identified in Garmisch relates to the notion that the tools and techniques used to develop computer software are inadequate for the complexity of the needed software. Consequently, much of the software engineering research conducted following this conference focuses on providing tools and techniques for reducing the uncertainty and assuring the quality of software with ever-growing complexity. Of a specific concern, is the uncertainty related to the requirements specifications phase. As the first phase in the development chain, the requirement specification phase has a profound effect on the quality of the entire software development process and on the final product, i.e., the hardware/software system.

One of the approaches for reducing uncertainty in the software development process, adopted by mainstream software engineering researchers, has been to introduce the discipline of quantitative software engineering. The quantitative software engineering research and practice stream, however, cannot completely cope with uncertainty, as some of this uncertainty is inherent to the process. Moreover, the fact that software engineering is a human-intensive process adds a challenging uncertainty dimension since human beings' reasoning is often characterized by inexact and fuzzy logic. This prompted a new interdisciplinary collaborative research direction that combines knowledge from the disciplines of uncertainty management and mitigation and the field of software engineering. Numerous research efforts in the area have been conducted, and many papers addressing soft computing and quantitative software engineering have been published [2–5]. Fuzzy logic is one of the most commonly and successfully used "tools" for handling uncertainty [6–10]. Indeed some papers addressing the role of fuzzy logic in quantitative software engineering have been published [11–14].

This chapter presents a new and innovative approach for fuzzy logic based quantitative software engineering procedures. The proposed complex fuzzy logic based model enables reasoning about processes with multi-dimensional components where each component is carrying fuzzy information and the interaction between the components cannot be decomposed and represented via primitive, one dimensional, fuzzy set theory and fuzzy logic operations such as conjunction, disjunction, negation, union, and intersection. In specific, we present the foundations of a complex fuzzy logic based inference system used to account for the intricate relations between software engineering constraints such as quality, software features, and development effort. The new model concentrates on the requirements specifications part of the software engineering process. Our model significantly improves the expressive power and inference capability of classic fuzzy logic as the tool for handling the uncertainty in this environment.

The problem addressed in this chapter boils down to the suitability of fuzzy logic as a soft computing model for dealing with uncertainty in software requirements specifications in tandem with applying quantitative software engineering methods. We ascertain that the fuzzy logic approach is a strong and excellent methodology for handling the uncertainty that is inherent in quantitative software engineering. Nevertheless, we show that the traditional single dimension fuzzy logic might fall short on dealing with real-world problems where several features such as quality, cost, development time, and usability, are involved. Especially, when these features are intertwined in a way that cannot be readily reduced to traditional fuzzy logic expressions composed of basic fuzzy logic connectives (conjunction, disjunction, negation, etc.).

The solution proposed is to use complex fuzzy logic as the underlying theory for dealing with the uncertainty involved in software requirements specifications. Via constructive examples we show that complex fuzzy logic is highly suitable for the task at hand.

The main contribution of the research described in the chapter is the formulation of a model that can enable better handling of the uncertainty in quantitative software engineering. To the best of our knowledge, this is the first research that is exploring the utility of complex fuzzy logic for handling uncertainty in the framework of quantitative software engineering. Furthermore, the research, which concentrates on software requirements specifications, can be extended to other phases of the software development process.

The rest of the chapter is organized in the following way. Sections 2 and 3, respectively, provide background concerning uncertainty involved in the software development process and the Quality Function Deployment approach to software requirements specifications. Section 4 contains a literature review listing relevant work. Section 5 introduces the concept of complex fuzzy logic and presents several ways in which it can be used for inference in the context of uncertainty in quantitative software requirements specifications. Finally, Sect. 6 includes the conclusions.

2 Uncertainty and the Software Development Process

One of the first techniques addressing the uncertainty and the growing complexity in the software development process was the Waterfall software development model introduced at the IEEE WESCON conference in 1970 by Royce [15]. In this seminal paper on the software development process; Royce introduced the first formalization of the process; organizing it into a series of five major sub-processes of: Requirements, Analysis, Design, Coding, and Testing. Over the years, researchers have made a number of changes to the model. One of the most significant changes was the absorption of the Analysis phase into the Requirements and Design phases. With the addition of minor name changes, the Waterfall development model has evolved to include four phases that are generally associated with the model: Requirements, Design, Implementation or Construction, and Validation.

Even though the Waterfall model has served the software development industry well for almost 50 years, there are a number of problems with the model. One of the most significant issues is that it focuses on reducing machine utilization with a result of increased personnel utilization, thereby making software development a labor-intensive process. Another issue with the Waterfall model is that the software developers are dependent on the quality of the requirement specifications established in the first phase of the process when analysts and developers know the least about the application. Frequently, these requirements originate from sources that do not understand the information necessary to build software and have a limited knowledge of the application, causing the developers to have questions about the requirement specifications. One of the major causes of costly software maintenance or project failure is poor requirement specifications [16, 17]. Hence, as developers began to gain more experience with the Waterfall model, they started to investigate a number of techniques to resolve the amount of human labor necessary to produce high quality software products and to reduce the impact of vague or incomplete software requirement specifications.

One of the first proposed approaches to address the uncertainty of software requirements was prototyping. Software prototypes can have two forms: throwaway and evolutionary [18]. A throwaway prototype provides information about the general structure and layout of the software but does not provide any information about the operation. A major disadvantage of this approach is that at the end of the design phase developers discard the prototype. Although this approach provides a great deal of information about user interfaces and links, it is expensive; and generally, it is not popular with the financial stakeholders within an organization. As its name implies, an evolutionary prototype is a working model of the desired software implemented without the use of traditional quality control tools. An evolutionary prototype becomes version-0, and the test engineers have the task of assuring that there are no defects in the software. From this version-0, analysts reverse engineer the software to create any required documentation.

Because of the many issues with prototyping, software engineers have turned to the notion of iterative software development. Iterative software development is a maintenance-based strategy used to reduce both risk and uncertainty during the construction of the application. One of the most widely known iterative techniques is Barry Boehm's Spiral Model [19]. In addition to the Spiral Model, most of the agile development methods also employ this concept for the same reasons [20–22].

In addition to the risks and uncertainty that are inherent in developing software, there is a great deal of uncertainty in describing the features needed in the software. Extracting the user needs and describing these needs in a format understandable by non-technical and technical individuals provides a source of considerable uncertainty in software engineering. Two of the major sources of uncertainty in the process of establishing the specifications for the needed software are the software engineers and the non-technical individuals providing the information upon which to base these specifications. Since it is unlikely that researchers will resolve all of the challenges in human communications any time soon, it is probably better to defer this challenge for future research. One of the tools being proposed as a first step in addressing these communication challenges is changing the perspective of requirement specifications to focusing on the tasks that the software's operators perform. A second tool proposed is utilizing soft computing methodologies for handling this uncertainty. Fuzzy logic has been successfully applied to resolve uncertainty at each of the five major processes of the waterfall model [23–26]. In this chapter, however, the soft computing model proposed is complex fuzzy logic.

3 Requirements Specification via Quality Function Deployment

In the 1970s, requirements engineers began to formulate a notion of the information that is necessary to develop a software application. These efforts evolved into the development of the IEEE Recommended Practice for Software Requirements Specifications, which is divided into sections describing the required interface, software functionality, non-functional or quality requirements, and constraints [16, 17]. The IEEE Recommended Practice-model centers on the items that are necessary for the software engineers to build the software. Although this general model has served the software industry very well, it does not provide a view of the software from the end user's perspective. The IEEE recommendation views security and usability as quality issues and documents them as non-functional requirements. This may explain the reason that these areas have remained challenges for software developers. Documenting software requirements from the perspective of the end-user or software operator (a.k.a. human centric) is an approach that is gaining in popularity. A human-centric approach to eliciting and documenting software requirements concentrates on the tasks that the software must support and on the operators that perform those tasks [27–29]. Generalizing this notion of viewing software from an operator's perspective yields the concepts of addressing software requirements from the external tasks (performed by operators and/or machines) that the software is intended to support. The Unified Modeling Language (UML) implements this concept in its use case diagram [30]. Software modeling techniques are also evolving to support the change to a user-centric approach [30–32]. One of the changes to software modeling techniques is the practice of employing use cases or user stories to describe the high-level characteristics of an application.

All of these innovations have acted to reduce the uncertainty of defining specifications and developing software applications, but there are many areas where the opportunity to further reduce challenges in software development activity exists. One such area is selecting the order of implementing the software requirements.

Selecting the order of implementing requirements can permit an early deployment of the product or service.

A number of innovative techniques have migrated into modern software development practice from research, conducted by Japanese investigators in the 1960s and 1970s, into improving manufacturing and quality assurance. One of the techniques making the migration from quality assurance into software development is Quality Function Deployment (QFD). Mizuno and Akao conducted research directed at bringing quality assurance into the design phase, rather than in the manufacturing phase resulting in QFD [33]. Their vision was to include the customers' view of the quality into all aspects of product design and manufacturing, thereby increasing the acceptance of the product in the marketplace [34].

According to Zultner, applying QFD to software requirements is a relatively simple process [35]. Customers receive a copy of the specification of an application and they assign one of three QFD categories to each requirement. These classification categories are *normal*, *expected*, and *exciting*. A customer classifying a requirement as normal means that a product such as the one specified **has** that feature. A requirement classified as expected means that the customer believes a product that does not contain that feature is disappointing. A requirement receiving an exciting classification is one exceeding what the customer expects to find in the specified product. There are two challenges not addressed in Zultner's discussion on QFD: Customer priority assignment and using QFD throughout the development process [35].

After receiving the customers' individual classification of the requirements, requirements engineers have several methods for establishing the classification of each requirement. One of these methods of applying customer priorities is to assign the requirement classification receiving the most votes. Another approach is to record the votes for each category providing later processes with more data for decision-making.

Although simple, QFD presents a number of challenges to requirements engineers. One important challenge that requirements engineers are facing is selecting the customers for providing the classifications because the quality of data is dependent on the customers' knowledge of the product and/or market. Another significant challenge for QFD relates to the quality of the customers asked to classify requirements. A QFD classification does not provide the requirements engineer insight into missing requirements. An advantage of QFD, outweighing both of these challenges, is that it provides the requirements engineer with customer insight as to the value of the specified facilities.

A challenge related to the customer's skill in evaluating requirements, but one that requirements engineers can control, is the focus and structure of the requirements. There are three major sections in a traditional requirement specification: interface, functional, and non-functional requirements [16, 17]. An issue arising with this type of document is that the interface, function and performance specifications are in three different locations making it difficult to pull all of this information together and classify each of the features. An approach that can improve requirement classification accuracy is an external-task or user-centric specification

[28, 30, 31]. A user-centric specification differs from traditional specification in that the requirements are organized based on the task that a customer is intended to perform with the software. Expressing the software's functionality in terms of use cases or user stories based on tasks they will perform with the software can produce better classifications [30, 31].

One of the possible uses, later in the development process, for these priorities is establishing implementation priorities. Prioritizing feature implementation is a significant challenge facing software developers using iterative development techniques such as Boehem's Spiral model or Schwalbe's Agile Scrum [19, 21]. In both the spiral model and Agile Scrum, developers must select a set of features for implementation during the next iteration. Usually, developers accomplish this by selecting features, based on effort estimates, fitting to the duration of the development increment. A better selection approach employs both the QFD classification and estimated effort. Using these two factors, is even more appropriate for situations where one or more iterations results in phase deployment or release.

Based on the definition of the QFD categories, it is apparent that it is an ordinal scale where requirements in an expected category are more desirable than requirements in the normal category; and requirements in the exciting category are more desirable than requirements in the expected category. Using this scale for development priorities would mean that exciting requirements are developed first followed by expected and then normal; but to have each iteration possess the maximum desirability to the customer base, the development priorities are expected requirements, followed by normal requirements, and then exciting. Implementing expected requirements is critical because they are the requirements that can increase customer dissatisfaction with the deployed product. Exciting requirements can increase the marketability of the product but might not improve customer satisfaction. Therefore, implementing the normal requirements before implementing the exciting requirements increases customer satisfaction and assures that the product is equal to the completion.

Designing the development process to work only on the requirements in a specific category is not a guarantee that developers will produce software maximizing the development time and overall customer satisfaction. Because of the complexity of the variables, traditional algorithmic approaches are not viable. Hence, a new approach to produce a list of requirements in rank order for an iteration cycle is needed.

The discussion in Sect. 2, has presented the software development process and the uncertainty involved in the process. The current section concentrated on using quantifiable methodologies for software specifications. It is quite clear that the quantification process reduces yet does not eliminate uncertainty. When it is all said and done, the engineers and stakeholders have to make decisions that optimize a utility, effort, and risk function. This function, however, is "ill defined" due to the inherent uncertainty and the fuzzy nature of human communication and human reasoning. For example, assigning ranks such as normal, expected, and exciting is a classic example of (human) fuzzy logic based reasoning. In this chapter, we propose to formulize two of the dimensions of the QFD space, namely utility and effort, using complex fuzzy logic. Later on, risk can be added as a third dimension in a multi-dimensional complex fuzzy logic based QFD process. In the next section, we list relevant work.

4 Literature Review

This section includes a review of literature associated with software requirements and describes work related to the use of fuzzy logic in formulating methods for handling uncertainty in software development. A new and innovative method for handling the uncertainty, which is proposed in this chapter, is the utilization of complex fuzzy logic. This original method is further elaborated in the next section.

A review of recent literature for software requirements reveals a limited amount of investigation into ways for writing and organizing requirements. Books like Wiegers' *Software Requirements*, Lauesen's *Software Requirements: Style and Techniques*, and Leffingwell's *Agile Software Requirements* discuss most of the research into writing and organizing requirements [36–38]. Each of these texts investigates most of the core issues of requirements analysis, but they do not investigate using formal methods for dealing with the uncertainty inherent in the process. Although similar, each text presents the topic from differing perspectives: traditional, linguistic, and lean software development methodologies, such as Agile-Scrum.

In the book *Software Requirements*, Weigers investigates most of the issues relating to the development of traditional requirements specification documents and the management of those requirements throughout the development process [36]. One of the features that make this book an important resource for the topic of software requirements is that it provides a large number of examples on soliciting requirements in a business environment. Even though Weigers addresses almost every aspect of software requirements, some might argue that the areas of specification style and Agile requirements practice need additional investigation. In the chapter addressing writing software requirements, Wiegers provides an excellent discussion on the mechanics of writing specifications, but he does not discuss the effects of different styles. In the book *Software Requirements: Styles and Techniques*, the author provides a better discussion on this issue. On lean software development or Agile methodologies [37], the discussion explains some of the differences between traditional requirements elicitation and the approach introduced with Agile-Scrum, but does not address the way that these differences affect the developers and the stakeholders. In the book *Agile Software Requirements*, Leffingwell provides a view of the effects of requirements on the developers and stakeholders [38].

The book *Software Requirements: Styles and Techniques* by Lauesen provides an overview to the requirements elicitation process, but focuses on linguistics techniques for achieving a specific objective [37]. Like Wiegers' approach, Lauesen provides a large number of cases studies and examples in writing requirements to achieve specific results and illustrates that different writing styles can achieve different results. This work, however, does not address ways for writing and organizing the requirements in order to enable software development using an Agile development methodology.

One of the most unusual approaches to software requirements specifications is described in the book *Agile Software Requirements* by Leffingwell [38]. In this work, Leffingwell combines Agile Modeling with requirements analysis and describes the ways that requirements are used in Agile development methodologies. The book suggests that requirements have a hierarchical characteristic, which is a subtle change from the "flat" approach suggested in other works. Using a hierarchical approach provides a level of details that is appropriate to the stakeholder and the developer.

One of the deficiencies that almost all of the works on software requirements have in common is their way of treatment of non-functional requirements, a.k.a. Quality Requirements or "ileitis". Originally, non-functional requirements were addressing system level topics such as reliability and maintainability. Over time, other topics such as human factors and security were introduced under non-functional requirements because many experts viewed these topics as system level issues that did not directly relate to the functionality of the software. Today two of the most severe challenges to software engineers are software usability and security.

In recent years, there has been a significant interest in the area of quantitative software engineering [2–5]. Several papers have addressed computational intelligence and quantitative software engineering [11–14]. Additionally, several survey papers and books/book-chapters such as [39–43] are useful in gaining access into recent developments in the field.

Alongside the interest in the general area of computational intelligence and software engineering, there has been increasing interest in the use of fuzzy set theory and fuzzy logic based reasoning as the soft computing paradigm [44–49]. With this respect [44, 45] are some of the most comprehensive accounts on fuzzy logic models in quantitative software engineering. The utilization of fuzzy logic to quantitative software engineering makes a lot of sense and provides highly valuable and usable tools for coping with the uncertainty in quantitative software engineering [44–49]. Nevertheless, this approach falls short of providing a rich and expressive way to take into account the intricate relations between major parameters affecting the software development process, such as quality, usability, development effort, and features included in release, cost, reliability, and risk. It is our assertion that the intricate relations can be effectively addressed using complex fuzzy logic.

Complex fuzzy logic has been introduced by Ramot et al. [50, 51] and several related applications have been considered [52]. Tamir et al. [53–56] refined the definition provided by Ramot and introduced examples where the interpretation provides for a rich and effective paradigm for reasoning which can capture uncertainty and human reasoning in a highly effective way.

An exhaustive search in research databases did not reveal any work that connects complex fuzzy logic with quantitative software engineering. To the best of our knowledge, this is the first research effort that reports on such a research direction.

5   Complex Fuzzy Systems

Several aspects of the software requirements specifications can utilize the concept of complex fuzzy logic [53]. Complex fuzzy logic can be used to represent the two-dimensional information embedded in the description of trade-offs between design effort and software feature inclusion. Additionally, complex fuzzy logic based inference can be utilized to exploit the fact that variables related to the uncertainty are inherent in the software requirements specifications. The software requirements space is multi-dimensional and cannot be readily defined via single dimensional clauses connected by single dimensional connectives. Finally, the multi-dimensional fuzzy space defined as a generalization of complex fuzzy logic can serve as a media for clustering of specifications related information in a linguistic variable-based feature space.

Tamir et al. [53, 55] introduced a new interpretation of complex fuzzy membership grade and derived the concept of pure complex fuzzy classes. This section includes a review of the concept of a pure complex fuzzy grade of membership, the interpretation of this concept as the denotation of a fuzzy class, and the basic operations on fuzzy classes.

To distinguish between classes, sets, and elements of a set we use the following notation: a class is denoted by an upper case Greek letter, a set is denoted by an upper case Latin letter, and a member of a set is denoted by a lower case Latin letter.

The Cartesian representation of the pure complex grade of membership is given in the following way:

$$\mu(V, z) = \mu_r(V) + j\mu_i(z),$$

where $\mu_r(V)$ and $\mu_i(z)$, the real and imaginary components of the pure complex fuzzy grade of membership, are real value fuzzy grades of membership. That is, $\mu_r(V)$ and $\mu_i(z)$ can get any value in the interval [0,1]. The polar representation of the pure complex grade of membership is given by:

$$\mu(V, x) = r(V)e^{j\sigma\phi(z)},$$

where $r(V)$ and $\phi(z)$, the amplitude and phase components of the pure complex fuzzy grade of membership, are real value fuzzy grades of membership. That is, they can get any value in the interval [0,1]. The scaling factor σ is in the interval $(0, 2\pi)$. It is used to control the behavior of the phase within the unit circle according to the specific application. Typical values of σ are $\{1, \frac{\pi}{2}, \pi, 2\pi\}$. Without loss of generality, for the rest of the discussion in this section we assume that $\sigma = 2\pi$.

The difference between pure complex fuzzy grades of membership and the complex fuzzy grade of membership proposed by Ramot et al. [50, 51] is that both components of the membership grade are fuzzy functions that convey information about a fuzzy set. This entails a different interpretation of the concept as well as a different set of operations and a different set of results obtained when these operations are applied to pure complex grades of membership. This is detailed in the following sections.

5.1 Complex Fuzzy Class

A fuzzy class is a finite or infinite collection of objects and fuzzy sets that can be defined in an unambiguous way and comply with the axioms of fuzzy sets given by Tamir et al. and the axioms of fuzzy classes given in [53, 54, 57, 58]. While a general fuzzy class can contain individual objects as well as fuzzy sets, a *pure fuzzy class of order one* can contain only fuzzy sets. In other words, individual objects cannot be members of a pure fuzzy class of Order 1. A pure fuzzy class of order M is a collection of pure fuzzy classes of order $M - 1$. We define a *Complex Fuzzy Class* Γ to be a pure fuzzy class of order one, i.e., a fuzzy set of fuzzy sets. That is, $\Gamma = \{V_i\}_{i=1}^{\infty}$; or $\Gamma = \{V_i\}_{i=1}^{N}$ where V_i is a fuzzy set and N is a finite integer. Note that despite the fact that we use the notation $\Gamma = \{V_i\}_{i=1}^{\infty}$, we do not imply that the set of sets $\{V_i\}$ is enumerable. The set of sets $\{V_i\}$ can be finite, countably infinite, or uncountably infinite. The use of the notation $\{V_i\}_{i=1}^{\infty}$ is just for convenience.

The class Γ is defined over a universe of discourse T. It is characterized by a pure complex membership function $\mu_\Gamma(V, z)$ that assigns a complex-valued grade of membership in Γ to any element $z \in U$ (where U is the universe of discourse). The values that $\mu_\Gamma(V, z)$ can receive lie within the unit square or the unit circle in the complex plane and are in one of the following forms:

$$\mu_\Gamma(V, z) = \mu_r(V) + j\mu_i(z),$$
$$\mu_\Gamma(z, V) = \mu_r(z) + j\mu_i(V),$$

where $\mu_r(\alpha)$ and $\mu_i(\alpha)$, are real functions with a range of [0,1].

Alternatively:

$$\mu_\Gamma(V, z) = r(V)e^{j\theta\phi(z)},$$
$$\mu_\Gamma(z, V) = r(z)e^{j\theta\phi(V)},$$

where $r(\alpha)$ and $\phi(\alpha)$, are real functions with a range of [0, 1] and $\theta \in (0, 2\pi]$.

In order to provide a concrete example, we define the following pure fuzzy class. Let the universe of discourse be the set of all the features that can be added to a specific software application along with a set of attributes related to the features, such as related development effort and perception of importance (i.e., "expected", "normal", and "exciting"). Let M_i denote the set of features considered in step i of the software development process. Furthermore, consider a function (f_1) that associates a number between 0 and 1 with each set of features, where this function reflects the level of importance of the features included in the set. In addition, consider a second function (f_2) that associates a number between 0 and 1 with each specific feature, where this function denotes the development effort associated with including the feature in step i of the software development process. The functions (f_1, f_2) can be used to define a pure fuzzy class of order 1. A compound of the two functions in the form of a complex number can represent the degree of membership in the pure fuzzy class of "highly desired features" for the set of features considered in the last k development steps.

Formally, let U be a universe of discourse and let 2^U be the power-set of U. Let f_1 be a function from 2^U to [0,1] and let f_2 be a function that maps elements of U to the interval [0,1]. For $V \in 2^U$ and $z \in U$ define $\mu_\Gamma(V, z)$ to be:

$$\mu_\Gamma(V, z) = \mu_r(V) + j\mu_i(z) = f_1(V) + jf_2(z)$$

Then, $\mu_\Gamma(V, z)$ defines a pure fuzzy class of order 1, where for every $V \in 2^U$, and for every $z \in U$, $\mu_\Gamma(V, z)$ is the degree of membership of z in V and the degree of membership of V in Γ. Hence, a complex fuzzy class Γ can be represented as the set of ordered triples: $\Gamma = \{V, z, \mu_\Gamma(V, z) | V \in 2^U, z \in U\}$

Depending on the form of $\mu_\Gamma(\alpha)$ (Cartesian or polar), $\mu_r(\alpha)$, $\mu_i(\alpha)$, $r(\alpha)$, and $\phi(\alpha)$ denote the degree of membership of z in V and/or the degree of membership of V in Γ. Without loss of generality, however, we assume that $\mu_r(\alpha)$ and $r(\alpha)$ denote the degree of membership of V in Γ for the Cartesian and the polar representations respectively. In addition, we assume that $\mu_i(\alpha)$ and $\phi(\alpha)$ denote the degree of membership of z in V for the Cartesian and the polar representations respectively. Throughout this chapter, the term *complex fuzzy class* refers to a pure fuzzy class with pure complex-valued membership function, while the term *fuzzy class* refers to a traditional fuzzy class such as the one defined by [57].

5.2 Degree of Membership of Order N

The traditional fuzzy grade of membership is a scalar defining a fuzzy set. It can be considered as degree of membership of order 1. The pure complex degree of membership defined in this chapter is a complex number that defines a pure fuzzy class. That is, a fuzzy set of fuzzy sets. This degree of membership can be considered as degree of membership of order 2 and the class defined can be considered as a pure fuzzy class of order 1. Additionally, one can consider the definition of a fuzzy set (a class of order 0) as a mapping into a one-dimensional space and the definition of a pure fuzzy class (a class of order 1) as a mapping into a two-dimensional space. Hence, it is possible to consider a degree of membership of order N as well as a mapping into an N-dimensional space. The following is a recursive definition of a fuzzy class of order N. Part 2 of the definition is not necessary; it is given in order to connect the term pure complex fuzzy grade of membership and the term grade of membership of order 2.

Definition

1. A fuzzy class of order 0 is a fuzzy set; it is characterized by a degree of membership of order 1 and a mapping into a one-dimensional space.
2. A fuzzy class of order 1 is a set of fuzzy sets. It is characterized by a pure complex degree of membership. Alternatively, it can be characterized by a degree of membership of order 2 and a mapping into a two-dimensional space.
3. A fuzzy class of order N is a fuzzy set of fuzzy classes of order $N - 1$; it is characterized by a degree of membership of order $N + 1$ and a mapping into an $(N + 1)$-dimensional space.

5.3 Generalized Complex Fuzzy Logic

A general form of a complex fuzzy proposition is: "$x...A...B...$" where A and B are values assigned to linguistic variables and "$...$" denotes natural language constants. A complex fuzzy proposition P can get any pair of truth values from the Cartesian interval $[0, 1] \times [0, 1]$ or from the unit circle. Formally a fuzzy interpretation of a complex fuzzy proposition P is an assignment of fuzzy truth value of the form $p_r + jp_i$, or of the form $r(p)e^{j\theta(p)}$, to P. In this case, assuming a proposition of the form "$x...A...B...$," then $p_{(r)}(r(p))$ is assigned to the term A and $p_i(\theta(p))$ is assigned to the term B.

For example, under one interpretation, the complex fuzzy truth value associated with the complex proposition:

"x is an expected yet highly difficult to implement feature of the application"

can be $0.1 + j0.5$. Alternatively, in another context, the same proposition can be interpreted as having the complex truth value $0.3e^{j0.2}$. As in the case of traditional propositional fuzzy logic, we use the tight relation between complex fuzzy classes / complex fuzzy membership to determine the interpretation of connectives. For example, let C denote the complex fuzzy set of "features that are exciting and easy to implement", and let $f_C = c_r + jc_i$, be a specific fuzzy membership function of C, then f_C can be used as the basis for the interpretation of P. Next we define several connectives along with their interpretation.

Table 1 Basic propositional fuzzy logic connectives

Operation	Interpretation
Negation	$f('P) = (1 + j1) - f(P)$
Disjunction	$f(P \oplus Q) = \max(p_R, q_R) + j \times \max(p_I, q_I)$
Conjunction	$f(P \otimes Q) = \min(p_R, q_R) + j \times \min(p_I, q_I)$
Implication	$f(P \rightarrow Q) = \min(1, 1 - p_R + q_R) + j \times \min(1, 1 - p_I + q_I)$

Table 1 includes a specific definition of connectives along with their interpretation. In this table P, Q and S denote complex fuzzy propositions and $f(s)$ denotes the complex fuzzy interpretation of S. We use the fuzzy Łukasiewicz logical system as the basis for the definitions [57, 59]. Hence, the max t-norm is used for conjunction and the min t-conorm is used for disjunction. Nevertheless, other logical systems, such as Gödel fuzzy systems, can be used [59, 60]

The same axioms used for fuzzy logic are used for complex fuzzy logic, and modus ponens is the rule of inference.

5.4 Complex Fuzzy Propositions and Connectives Examples

Consider the following propositions (P, Q, and S respectively):
P "x is a very **exciting** yet highly **difficult to implement** feature."
Q "x is **expected** yet quite **easy to implement** feature."
S "x is a high **ranked** feature planned for release in the near **future**."

Let A be the term "*x is an exciting feature*" and let B denote the term "*difficult to implement.*" Furthermore, let C be the term " *is an expected feature,*" let D be the term "*x is a high ranked feature,*" and let E be the term "*future.*" Hence, P is of the form: "x is a very A that is highly B," and Q is of the form "x is C that is not quite B." In this case, the terms "*expected,*" "*normal,*" "*difficult,*" "*ranked,*" and "*future*" are linguistic variables. Furthermore, a term such as "*exciting,*" can get fuzzy truth values (between 0 and 1) or fuzzy linguistic values such as "*moderately,*" "*highly,*" and "*very,*" (the terms "*is,*" "*that,*" etc. are linguistic constants). Assume that the complex fuzzy interpretation (i.e., degree of confidence or complex fuzzy truth value) of P is $p_r + jp_i$, while the complex fuzzy interpretation of Q is $q_r + jq_i$. Thus, the truth value of "*x is an exciting feature,*" is p_R, and the truth value assigned to "*x is difficult to implement,*" is p_i. The truth value of "*x is an expected feature,*" is q_r. Suppose that the term "*easy*" stands for "*not **difficult**,*" the term "*low,*" stands for "*not high,*" and the term "*dull*" stands for "*not **exciting***". In a similar way, S is of the form: "x is high D that is ... near E," where the complex fuzzy interpretation of S is $s_r + js_i$. This, however, is not the only way to define these linguistic terms, and it is used to exemplify the expressive power and the inference power of the logic. Hence, the complex fuzzy interpretation of the following composite proposition is:

1. $f('p) = (1 - p_r) + j(1 - p_l)$

   That is, $'P$ denotes the proposition "x is a dull yet easy to implement feature."
   The confidence level in $'P$ is $(1 - p_r) + j(1 - p_i)$, where the fuzzy truth value of
   the term "x is a non **exciting** feature," is $(1 - p_r)$ and the fuzzy truth value of the
   term "x is an **easily implemented** feature." is $(1 - p_i)$

2. $f(P \oplus Q) = \max(p_r, 1 - q_r) + j \times \max(p_i, 1 - q_i)$.

   That is, $(P \oplus Q)$ denotes the proposition "x is a very **exciting** yet highly
   difficult to implement feature." OR

   "x is an **expected** yet quite **easy to implement** feature." The truth values of
   individual terms, as well as the truth value of $P \oplus Q$ are calculated according to
   Table 1.

3. $f('P \otimes Q) = \min(1 - p_r, q_r) + j \times \min(1 - p_i, q_i)$

   That is, $('P \otimes Q)$ denotes the proposition "x is a **dull** yet **difficult to implement**
   feature." AND

   "x is an **expected** yet quite **easy to implement** feature." The truth values of
   individual terms, as well as the truth value of $'P \otimes Q$ are calculated according to
   Table 1.

4. Let the term R stand for $(P \oplus Q)$, (the complex fuzzy interpretation of R is
   $r_r + jr_i$.) then,

$$R \rightarrow S = \min(1, 1 - r_r + s_r) + j \times \min(1, 1 - r_i + s_i).$$

Thus, $(R \rightarrow S)$ denotes the proposition

IF "x is a very **exciting** yet highly **difficult to implement** feature." OR

"x is an **expected**, yet quite **easy to implement** feature."

THEN   ` x is a high **ranked** feature planned for release in the near **future**." The
truth values of individual terms, as well as the truth value of $R \rightarrow S$ are calculated
according to Table 1.

5.5   Complex Fuzzy Inference Example

Assume that the degree of confidence in the proposition R as defined above is
$r_r + jr_i$, and assume that the degree of confidence in the fuzzy implication $T =
R \rightarrow S$ is $t_r + jt_i$. Then, using modus ponens

R

$\dfrac{R \rightarrow S}{}$

S

One can infer S with a degree of confidence $\min(r_r, t_r) + j \times \min(r_i, t_i)$.

In other words if one is using:

"x is an **exciting** yet **difficult to implement** feature." OR "x is an **expected** yet **easy to implement** feature.

IF "x is an **exciting** yet **difficult to implement** feature." OR

"x is an **expected** yet **easy to implement** feature."

THEN "x is a high **ranked** feature planned for release in the near **future**."

"*x is* a high **ranked** feature planed for release in the near **future**."

Hence, using modus ponens one can infer:

"x is a high **ranked** feature planned for release in the near **future**" with a degree of confidence of $\min(r_r, t_r) + j \times \min(r_i, t_i)$.

This example shows the potential of complex fuzzy inference to enhance the ability for resolving uncertainty involving the requirements specifications process. The actual process of using this approach for inference is described [51]. In this case a complex fuzzy rule-based system is generated via complex fuzzification and used for complex fuzzy inference. Eventually via de-fuzzification actual crisp conclusions are obtained [51]. In [26] we have described Software Testing Using Artificial Neural Networks and Info-Fuzzy Networks. We are currently working on extending this research to using complex fuzzy inference. Finally, we are currently exploring the use of complex fuzzy logic and inference for non-functional requirements such as usability requirements.

6 Conclusions

In this chapter, we have introduced an innovative approach for fuzzy logic based quantitative software engineering procedures. We have presented a complex fuzzy logic based inference system used to account for the intricate relations between software engineering constraints such as quality, software features, and development effort. The model presented concentrates on the requirements specifications part of the software engineering process. Furthermore, the presented model significantly improves the expressive power and inference capability of the soft computing component in the soft computing based quantitative software engineering.

In the future, we plan to concentrate on software requirements for human computer interaction applications. Additionally, we plan to further investigate the utility of the new model in the development of software requirements for large-scale software systems. Furthermore, we plan to increase the dimensionality of the fuzzy terms to include other factors such as risk, reliability, usability etc. Finally, we plan to expand the work to include other components of the software development process.

References

1. Software Engineering: Report of a Conference Sponsored by the NATO Science Committee, Garmisch, Germany, 7–11 Oct 1968. Scientific Affairs Division, NATO, Brussels (1969)
2. Kim, M., Zimmermann, T., Nagappan, N.: An empirical study of refactoring challenges and benefits at microsoft. Softw. Eng. IEEE Trans. **40**(7), 633–649 (2014)
3. Lazaro, M., Marcos, E.: An approach to the integration of qualitative and quantitative research methods in software engineering research. In: CAISE*06 Workshop on Philosophical Foundations on Information Systems Engineering, Luxemburg (2006)
4. Dyba, T.: An empirical investigation of the key factors for success in software process improvement. Softw. Eng. IEEE Trans. **31**(5), 410–424 (2005)
5. Verner, J.M., Evanco, W.M.: In-house software development: what project management practices lead to success? IEEE Softw. **22**(1), 338–353 (2005)
6. Zadeh, L.A.: Fuzzy sets. Inf. Control **8**, 338–353 (1965)
7. Zadeh, L.A.: Fuzzy sets as a basis for a theory of possibility. Fuzzy Sets Syst. **1**, 3–28 (1978)
8. Kandel, A.: Fuzzy mathematical techniques with applications, reading. Addison-Wesley, MA (1986)
9. Tamir, D.E., Kandel, A.: An axiomatic approach to fuzzy set theory. Inf. Sci. **52**, 75–83 (1990)
10. Tamir, D.E., Kandel, A.: Fuzzy semantic analysis and formal specification of conceptual knowledge. Info. Sci. Intell. Syst. **82**(3–4), 181–196 (1995)
11. Pillai, S.K., Jeyakumar, M.K.: Evaluation of neural networks for software development effort estimation using a new criterion. SigSoft Softw. Eng. Notes **39**(5), 1–6 (2014)
12. Nagendra Kumar, G., Aswani Kumar, C.: Generation of high level views in reverse engineering using formal concept analysis. In: First International Conference on Networks & Soft Computing (ICNSC), Hyderabad (2014)
13. Singh, C., Pratap, A., Singhal, A.: Estimation of software reusability for component-based systems using soft computing techniques. In: Confluence The Next Generation Information Technology Summit (Confluence), 2014 5th International Conference, Noida (2014)
14. Bakshi, T., Sarkar, B., Sanyal, S.K.: A new soft-computing based framework for project management using game theory. In: 2012 International Conference on Communications, Devices and Intelligent Systems (CODIS), Kolkata (2012)
15. Royce, W.W.: Managing the development of large software systems: concepts and techniques. In: IEEE WESON 26 (August): 1–9, Los Angeles (1970)
16. IEEE: IEEE Std 1233-1998 IEEE Guide for Developing System Requirement Specification. IEEE Computer Society, New York, NY (1998)
17. IEEE: IEEE Std-830-1988 IEEE Recommended Practice for Software Requirements. IEEE Computer Society, New York, NY (1998)
18. Pressman, R.: Software Engineering: A Practioner's Approach. McGraw-Hill, New York, NY (2010)
19. Boehm, B.W.: A spiral model of software development and enhancement. Computer **21**(5), 61–72 (1988)
20. Ambler, S.W.: Agile modeling. Ambysoft Inc. (2014). [Online]. Available: http://www.agilemodeling.com. Accessed 30 Sept 2014
21. Schwaber, K., Beedle, M.: Agile Software Development with Scrum. Prentice-Hall, Upper Saddle River, NJ (2002)
22. Beck, K., Andres, C.: Extreme Programming Explained: Embrace Change, 2nd edn. Addion-Wesley Professional (2004)
23. Pedrycz, W., Succi, G., Reformat, M., Musilek, P., Bai, X.: Self-organizing maps as a tool for software analysis. In: Canadian Conference of Electrical and Computer Engineering 2001, Toronto, Canada (2001)
24. Noppen, J., van den Broek, P., Aksit, M.: Dealing with fuzzy information in software design methods. In: Proceedings of the IEEE Annual Meeting of the North American Fuzzy Information Processing NAFIPS '04, Alberta, Canada (2004)

25. Achimugu, P., Selamat, A., Ibrahim, R., Mahrin, M.: An adaptive fuzzy decision matrix model for software requirements prioritization. In: Sobecki, J.B.V.C.S. (ed.) Advanced Approaches to Intelligent Information and Database Systems, vol. 551, pp. 129–138. Springer International Publishing, Switzerland (2014)
26. Agrawal, D., Tamir, D.E., Last, M., Kandel, A.: A comparative study of software testing using artificial neural networks and info-fuzzy networks. IEEE Trans Man Mach. Cybern. **42**(5), 1183–1193 (2012)
27. Hickey, A.M., Davis, A.M.: A unified model of requirements elicitation. J. Manage. Info. Syst. **20**(4), 65–84 (2004)
28. Constantine. L.L., Lockwood, L.A.D.: Software for Use: A practical Guide to the models and methods of Usage-Centered Design, Reading. ACM Press, MA (1999)
29. Sutcliffe, A.: Scenario-based requirments analysis. Requirements Eng. **3**(1), 48–65 (1998)
30. Booch, G., Rumbaugh, J., Jacobon, I.: Unified Modeling Language User Guide, 2nd edn. Addison-Wesley Professional, New York, NY (2005)
31. Ambler, S.W.: Agile Modeling: Effective Practices of Modeling and Documentation. Ambysoft Inc. (2014). [Online]. Available: http://www.agilemodeling.com/. Accessed 11 Nov 2014
32. Jacobson, I.: Object Oriented Software Engineering: A Use Case Driven Approach, Redwood City. Addison Wesley Longman Publishing Co., CA, USA (2004)
33. Mazur, G.: History of QFD. Quality Function Deployment Institute [Online]. Available: http://ww.gfdi.org/what_is_qfd/history_of_qfd.html. Accessed 29 Sept 2014
34. Akao, Y.: Quality Function Deployment: Integrating Customer Requirements into Product Design. Productivity Press, Cambridge, MA (1990)
35. Zultner, R.E.: Quality function deployment for software: satisfying customers. American Programmer, pp. 28–41 (1992)
36. Wiegers, K., Beatty, J.: Software Requirements. Microsoft Press, Readmond, WA (2013)
37. Lauesen, S.: Software Requirements: Styles and Techniques. Pearson Education, Edinburg Gate (2002)
38. Leffingwell, D.: Agile Software Requirements: Lean Requirements Practices for Teams, Programs and the Enterprise. Pearson Education Inc., Boston, MA (2011)
39. Bhuyan, M.K., Mohapatra, D.P., Sethi, S.: A survey of computational intelligence approaches for software reliability prediction. ACM SigSoft Softw. Eng. Notes **39**(2), 1–10 (2014)
40. Harman, M., Mansouri, S.A., Zhang, Y.: Search-based software engineering: trends, techniques and applications. ACM Comput. Surv. **45**(1), 61 (2012)
41. Lee, J.: Software Engineering with Computational Intelligence. Springer, Secaucus, NJ (2003)
42. Dick, S., Kandel, A.: Computational Intelligence in Software Quality Assurance. World Scientific, Singapore (2005)
43. Last, M., Kandel, A.: Automated test reduction using an info-fuzzy network. In: Software Engineering with Computational Intelligence, vol. 731, pp. 235–258. Springer International (2003)
44. Pedrycz, W., Succi, G.: Fuzzy logic classifiers and models in quantitative software engineering. In: Advances in Machine Learning Applications in Software Engineering, pp. 146–167. IGI Global, USA (2007)
45. Pedrycz, W., Breuer, A., Pizzi, N.J.: Fuzzy adaptive logic networks as hybrid models of quantitative software engineering. Intell. Autom. Soft Comput. **12**(2), 189–209 (2008)
46. Lee, J., Kuo, J.-Y.: New approach to requirements trade-off analysis for complex systems. Knowl. Data Eng. IEEE Trans. **10**(4), 551–562 (1998)
47. Georgieva, O., Dimov, A.: Software reliability assessment via fuzzy logic model. In: Proceedings of the 12th International Conference on Computer Systems and Technologies (CompSysTech '11), Vienna, Austria (2011)
48. Cooper, K., Cangussu, J.W., Lin, R., Sankaranarayanan, G., Soundararadjane, R., Wong, E.: An empirical study on the specification and selection of components using fuzzy logic. In: Proceedings of the 8th international conference on Component-Based Software Engineering (CBSE'05), St. Louis (2005)

49. George, R., Srikanth, R., Petry, F.E., Buckles, B.P.: Uncertainty management issues in the object-oriented data model. Fuzzy Syst. IEEE Trans. **4**(2), 179–192 (1996)
50. Ramot, D., Milo, R., Friedman, M., Kandel, A.: Complex Fuzzy Sets. Fuzzy Syst. IEEE Trans. **10**(2), 171–186 (2002)
51. Ramot, D., Friedman, M., Langholz, G., Kandel, A.: Complex fuzzy logic. Fuzzy Syst. IEEE Trans. **11**(4), 450–461 (2003)
52. Dick, S.: Towards complex fuzzy logic. Fuzzy Syst. IEEE Trans. **13**(3), 405–414 (2005)
53. Tamir, D.E., Lin, J., Kandel, A.: A new interpretation of complex membership grade. Int. J. Intell. Syst. **26**(4) (2011)
54. Tamir, D.E., Kandel, A.: Axiomatic theory of complex fuzzy logic and complex fuzzy classes. Int. J. Comput. Commun. Control **6**(3) (2011)
55. Tamir, D.E., Last, M., Kandel, A.: Complex fuzzy logic. In: Seising, R., Trillas, E., Termini, S., Moraga, C. (eds) On Fuzziness, p. 429. Springer, Heidelberg (2013)
56. Tamir, D.E., Last, M., Kandel, A.: The theory and applications of generalized complex fuzzy propositional logic. In: Yager, R.R., Abbasov, A.M., Reformat, M.Z., Shahbazova, S.N. (eds) Soft Computing: State of the Art Theory and Novel Applications. Springer, Heidelberg (2013)
57. Behounek, L., Cintula, P.: Fuzzy class theory. Fuzzy Sets Syst. **154**(1), 34–55 (2005)
58. Fraenkel, A.A., Bar-Hillel, Y., Levy, A.: Foundations of Set Theory, 2nd edn. Elsivier, Amsterdam (1973)
59. Cintula, P.: Advances in LΠ and LΠ1/2 logics. Arch. Math. Logic **42**, 449–468 (2003)
60. Montagna, F.: On the predicate logics of continuous t-norm BL-algebras. Arch. Math. Logic **44**, 97–114 (2005)

Author Biographies

Dr. Tamir is an associate professor in the Department of Computer Science, Texas State University, San Marcos, Texas (2005—to date). He obtained the Ph.D.-CS from Florida State University in 1989, and the M.S./B.S.-EE from Ben-Gurion University, Israel in 1983, 1986 respectively.

From 1996–2005, he managed applied research and design in DSP Core technology in Motorola-SPS/Freescale. From 1989–1996, he served as an assistant/associate professor in the CS Department at Florida Tech. Between 1983 and 1986, he worked in the applied research division, Tadiran, Israel.

Dr. Tamir is conducting research in the areas of data compression and pattern recognition, complex fuzzy logic, and effort based usability evaluation. Additional research areas include signal processing and combinatorial optimization. He is teaching graduate and undergraduate courses in formal languages, computer architecture, multi-media programming, graphical user interfaces, and computer graphics. He is supervising research and individual study courses with graduate and undergraduate students; twenty eight students have completed their master's thesis/Ph.D. dissertation under his supervision.

Dr. Tamir has published more than 90 refereed journal and conference papers as well as four book chapters in the areas of combinatorial optimization, computer vision, audio, image, and video compression, human computer interaction, and pattern recognition. He has been a member of the Israeli delegation to the MPEG committee and a Summer Fellow at NASA KSC.

Dr. Mueller is an assistant professor in the Department of Computer Information Systems, Texas A&M University—Central Texas, Killeen, Texas (2012—to date). He obtained the M.S. and Ph.D. degrees in Computer Science from the Illinois Institute of Technology in 1997 and 2003 respectively, and the B. T. degree in Computer Applications from Washington University in St. Louis Missouri in 1975.

From 1968 to 1985, he acted in various capacities developing and testing inventory control software, data base query languages, and telephony applications. From 1985 to 2008, he was an independent-contract Software Engineer. He has developed and tested high reliability software such as medical devices, specialized networking application, and avionics.

Dr. Mueller is a Senior Member of the ACM and the IEEE. In addition, he is a member of $(ISC)^2$ holding the Certified Information Systems Security Professional (CISSP) credential.

Dr. Mueller is conducting research in several Software Engineering topics including cyber-security, software requirements, software testing, and effort based software usability. He has supervised 3 master's thesis and published 15 refereed journal and conference papers on human factors, cyber-security, and software testing.

Dr. Kandel received the B.S. degree in Electrical Engineering from the Technion-Israel Institute of Technology, Haifa, Israel, the M.S. degree in Electrical Engineering from the University of California, Santa Barbara, and the Ph.D. degree in Electrical Engineering and Computer Science from the University of New Mexico, Albuquerque.

He was the Chairman of the Computer Science and Engineering Department, University of South Florida, during 1991–2003 and the Founding Chairman of the Computer Science Department at the Florida State University during 1978–1991. He was a Distinguished University Research Professor and Endowed Eminent Scholar in Computer Science and Engineering at the University of South Florida, Tampa, and the Executive Director of the National Institute for Applied Computational Intelligence.

Dr. Kandel is the author or coauthor of more than 800 research papers for numerous professional publications in Computer Science and Engineering. He is the author, coauthor, editor, or coeditor of 51 text books and research monographs, with the most recent text "Calculus Light" coauthored with Prof. M. Friedman and published in 2011 by Springer-Verlag. He is a member of the editorial boards and advisory boards of several leading international journals in Computer Science and Engineering.

Dr. Kandel is a Fellow of the Association for Computing Machinery, the Institute for Electrical and Electronics Engineers, the New York Academy of Sciences, the American Association for the Advancement of Science, and the International Fuzzy Systems Association. He was the recipient of the Fulbright Senior Research Fellow Award at Tel-Aviv University during 2003–2004. In 2005 and 2013, he was selected by the Fulbright Foundation as a Fulbright Senior Specialist in applied fuzzy logic and computational intelligence. Dr. Kandel is the recipient of numerous professional awards, including the 2012 IEEE Computational Intelligence Society Fuzzy Systems Pioneer Award.

Presently, Dr. Kandel is a Distinguished University Professor Emeritus at the University of South Florida (retired in 2012), and since 2013 he is a Visiting Professor in the School of Computing and Information Sciences at the Florida International University, Miami, Florida.

Possibilistic Assessment of Process-Related Disclosure Risks on the Cloud

Valerio Bellandi, Stelvio Cimato, Ernesto Damiani and Gabriele Gianini

Keywords Cloud computing · Risk assessment · Secure computation · Possibility · Theory

1 Introduction

Business processes that involve storing or transmitting personal data are subject to strict regulatory and compliance requirements. The choice of deploying such processes on a shared platform like the cloud hinges on the process owner being convinced that the cloud platform is fully compliant with regulations. If a highly regulated business process (e.g. a e-health or e-government one) is to take place on a public cloud, then its deployment must fully meet all applicable regulations and laws regarding data confidentiality and leakage prevention. Otherwise, the process owner risks liability for violating privacy or other legal requirements. In this chapter we focus on a specific but important category of data disclosure events, the ones that bring one or more parties taking part to a cloud-based business process to know more information than the process execution would entail. These unwanted disclosures may be due to intentional publishing of supposedly protected information items, or to carelessness in the communication protocol implementation and deployment, e.g. when one party is using the same mobile terminal previously used by another and can reconstruct the information items held. We call these events

V. Bellandi (✉) · S. Cimato · E. Damiani · G. Gianini
Department of Computer Science, Università degli Studi di Milano, Milano, Italy
e-mail: valerio.bellandi@unimi.it

S. Cimato
e-mail: stelvio.cimato@unimi.it

E. Damiani
e-mail: ernesto.damiani@unimi.it

G. Gianini
e-mail: gabriele.gianini@unimi.it

© Springer International Publishing Switzerland 2016
W. Pedrycz et al. (eds.), *Computational Intelligence and Quantitative Software Engineering*, Studies in Computational Intelligence 617,
DOI 10.1007/978-3-319-25964-2_9

process-related data disclosures, in order to distinguish them from disclosures due
to conventional eavesdropping attacks. Here we present a novel, *process-oriented*
risk assessment methodology aimed at assessing process-related data disclosure
risks on cloud computing platforms. likelihood estimates, which allow analysing
threats for which a detailed history is not available (iii) support for quick visual
comparisons of risk profiles from alternative processes even when impact cannot be
exactly quantified.

Risk Definition

From the economic standpoint, the *risk* of an "adverse event" (a.k.a. "feared event")
E for a given actor A is often represented as the product of the damage
$I(E)$ (expressed in currency units) in which A incurs when E really happens, times
the likelihood that E might happen, traditionally represented in terms of probabilities. In symbols: $R(A, E) = I(E) Pr(E)$.

In the computer security context, one needs to identify all the adverse events as
manifestations of security threats and for each estimate $I(E)$ and $Pr(E)$; then the
overall risk is computed by a suitable aggregator. Typically [1], the risk analyst puts
herself in the place of a specific actor (e.g. the *process owner*, i.e. the stakeholder in
whose name, a business process P is executed) and asks the following questions:

- Which adverse event can happen to the information assets involved in P? (*threat
  categorization*)
- How severe could that event be for the process owner? (*threat impact
  assessment*)
- How much is this event plausible? (*threat likelihood* assessment)

Likelihood

As far as the estimate of the likelihood is concerned, many risk assessment methods
use predictive models involving a certain number of parameters, and each parameter is usually affected by some degree of uncertainty. There are cases where the
probability of adverse events can eventually be obtained by propagating this
uncertainty (i.e. either by analytical or Monte Carlo methods). However, in many
cases, assessing the probability of the parameter values is too difficult due to
incompleteness of information. As put forward in [2] "*Probability is perfect, but we
cannot elicit it perfectly*". In fact, the uncertainty involved in risk assessment has
typically two distinct origins: it can be due either to *variability* or to *imprecision* (or
to both factors). *Variability* (also referred to as "objective uncertainty") arises often
(but not exclusively) from the *random* character of natural processes. *Imprecision*
(also referred to as incomplete information, partial ignorance or "subjective
uncertainty") arises from the *partial* character of *individual knowledge* about the
state of the world. Traditionally, in risk assessment no distinction was made
between these two types of uncertainty, both being represented by means of a single
probability distribution. However, in case of partial ignorance, the use of probability measures introduces information that is in fact not available: this may seri-
ously bias the outcome of a risk analysis [3]. While random variability can be

suitably represented by probability measures and propagated by the methods of *Probability Theory*, incomplete information is better accounted for and propagated by the methods of *Possibility Theory*. Possibility Theory is similar to Probability Theory in that it is based on set functions, but differs from it by the use of *a pair* of dual set-functions (the *possibility measure* and the *necessity measure*) instead of only one (the probability measure) and by the fact that is not additive, but sub-additive, and can be cast either in an ordinal or in a numerical setting (the two formulations differ in terms of conditioning, and independence notions). The latter form will be used in this chapter. The development of Possibility Theory is due to a large number of authors, especially to Zadeh [4] and to Dubois and Prade [5, 6] (a systematic exposition can be found in [7], a review in [8–10]). In our methodology, we rely on the knowledge of the business process model and its underlying micro-economics to attach possibilities to actors misbehavior/violation of confidentiality. We are going to indicate (Sect. 5.2) how the knowledge about the likelihood of an adverse event can be elicited and suitably represented in possibilistic form.

Impact

As far as the impact (a.k.a. severity) is concerned, its precise quantification is often a challenge. In our methodology, we provide an evaluation of costs taking into account the value of the disclosed information by means of a set of techniques known as Value of Information (VoI) analysis. As we shall see, uncertain knowledge about the impact of adverse events on processes can be suitably represented in possibilistic form (Sect. 5.3). By composing possibilistic information about the likelihood of an event and the possibilistic information about the impact of that event, one can obtain a possibilistic representation of the risk of an adverse event and use this as the basis for taking decisions.

In this chapter, first we review related work (Sect. 2); then we briefly introduce Possibility Theory (Sect. 3) and, after illustrating the structure of a generic process model (Sect. 4), we describe our methodology (Sect. 5). Finally, we define an instance of our process model and illustrate the application of our methodology to that specific case (Sect. 6).

2 Related Work

Cloud computing is a computing paradigm where "massively scalable IT-enabled capabilities are delivered 'as a service' to external customers using Internet technologies" [11]. If on the one hand the adoption of such a model provides cost savings through economies of scale, on the other it introduces some peculiar risk challenges that increase the risks traditionally introduced by any externally provided IT service.

Security Risk Assessment on the Cloud

From the perspective of the security analyst, cloud-based services are outsourced in the least transparent way, since data are stored and processed on unspecified servers located in some unknown places, out of the control of the data owner. For these reasons, some researchers have started introducing techniques to deal with specific cloud-related issues [12–14].

Various bodies such as the Cloud Security Alliance (CSA), the European Network and Information Security Agency (ENISA), and the US National Institute of Standards and Technology (NIST) have released documents assisting organizations and customers in the evaluation of the security issues related to cloud computing [15–17]. The Cloud Controls Matrix released by CSA provides an useful description of the security principles aiming to guide cloud vendors and help cloud clients in assessing overall security risks of a cloud service provider [16]. NIST Special Publication 800-144 provides an overview of the security and privacy challenges for public cloud computing. In an early study by ENISA [15], a cloud-specific, semi-qualitative risk assessment process was anecdotally described using three use-cases: the SMEs' perspective on cloud computing, the impact of cloud computing on service resilience, and a scenario on cloud-based e-Health applications. Within the feared events mentioned in the study is the *unwanted disclosure of information* to co-tenants: an event which may be due to failure of mechanisms separating storage, memory and routing between different tenants of the shared infrastructure caused by different kind of threats, such as *guest-hopping attacks*, or SQL injection attacks exposing multiple customers' data stored in the same table.

The industrial whitepaper [18] describes a qualitative risk assessment methodology specific for clouds. It starts by considering risk factors that change when an organization shifts from a traditional infrastructure to a cloud-based one. The analysis is based on the risk taxonomy presented by the Open Group [19]. Some cloud specific threats are identified, such as the possibility for an attacker to escape from the virtualized environment, the possibility to ride or hijack sessions in shared web applications, threats to the integrity and confidentiality of data caused by the insecure usage of cryptography or the selection of flawed implementation of cryptographic primitives. Other specific threats are those concerning problems with standard security controls, such as the difficulties to execute network security controls in virtualized environment, poor management of the of encryption keys, difficulty of establishing security metrics suitable to monitor the status of cloud resources.

Another interesting case study showing qualitative risk assessment at work in a cloud computing scenario is described in [20], where the case of a software company developing business software and adopting a IAAS provided by another CSP is analyzed. The methodology is based on the *Risk IT* framework, which provides a detailed process model for the management of IT-related risk, as well as on the *COBIT 5* framework by the Information Systems Audit and Control Association (ISACA) [21], which assists enterprises in achieving their objectives for the governance and management of Information Technology (IT). RISK IT includes a list of generic high-level risk scenarios and a mapping between those scenarios and more general COBIT control objectives, so that a map of risks showing the impact/magnitude and likelihood/frequency of key risks can be created. Based on this map organisations adopt a risk mitigation approach, balancing the benefit from deploying security controls and the costs necessary for their implementation.

Some initial work toward a quantitative risk assessment framework for cloud computing, called QUIRC, has been presented in [22]. The QUIRC framework classically defines risk as a combination of (a rough estimate of) the probability of a feared event and its severity. QUIRC lists six key *Security Objectives* (SO) for cloud platforms, claiming that most of the typical attack vectors and feared events map to one of these six categories. Its strong point is its fully quantitative approach, which enables stakeholders to comparatively assess the robustness of cloud vendor offerings. However, lack of reliable data on the occurrences of cloud threats in many vertical domains can make QUIRC probability assignment somewhat arbitrary.

Another quantitative framework for assessing some security risks associated with cloud computing platforms has been proposed in [23]. The model relies on subjective assessments by experts of likelihood and severity of adverse events that allows the definition of the weights of the coefficients for the basic security properties (CIA—Confidentiality, Integrity, and Availability) and the corresponding values of assets relevant for the project. Even if the resulting prioritisation of the risks is valuable, this approach only considers threats to some specific security properties.

Focusing on the same set of security properties, Khan et al. [24] introduced a systematic approach combining existing tools and techniques such as CORAS [25], and the IRAM (Information Risk Analysis Methodology) with the Threat and Vulnerability Assessment tool (T&VA) [26]. Their technique uses a list of threats provided by the Information Security Forum (ISF). Depending on the priority of the assets and on the perceived likelihood of the ISF threats, they construct an evaluation matrix and use it to rate the threats' impact on the business. Due to the anecdotal nature of the ISF threat list, this technique can be considered a semi-quantitative one.

Integration of Disclosure Risk Assessment with Privacy Risk Management Frameworks

Business processes involving personal data present specific risks due to the liability brought upon the process owner (often called *controller* in this context) and, possibly, upon other stakeholders by violations of the privacy of third parties (*data subjects*). A special regulatory framework for personal data processing is currently in force at the European level, prescribing—among other things—that the purposes of the business process involving personal data are clearly defined, that personal data are relevant to such purposes, that personal data are erased at the end of a given time, and that all data subjects have the opportunity to exercise their rights (such as opposition, access, rectification and deletion of their personal data). In addition, the controller has an obligation to take all useful precautions in order to ensure the security of the personal data he processes.

Privacy authorities and regulators have been devoting a huge effort to develop *Privacy Risk Management* (PRM) frameworks [27]. As observed in [28], we still lack of a systematic approach to identify privacy threats and design privacy supportive business processes. According to recent studies [29], privacy threat analysis should begin at the earliest possible stage of the lifecycle of any business process involving personal data, when there are more opportunities to influence the business process' implementation; also, it should continue along the business process lifecycle. In principle, privacy risks may be targeted by all forms of risk analysis introduced in previous Sections. *Qualitative Analysis* uses ordinal scales expressed in words to quickly assess the relative severity of risks. This technique is often used when numerical data is not available and/or the process targeted by privacy risk assessment is only partially known by the risk assessor, as is often the case at early design stage. The quantitative approach is generally more complex to undertake, requiring full knowledge of the business processes to be analyzed and, in many cases, the development of organization-specific value models to assess the value of disclosed information as seen by different actors.

Possibility Theory in Risk Assessment

The drawbacks of the approach in computing the risk as a product of perfectly known factors, such as probability and impact, have been pointed out by several authors. A major criticism regards the way experts assign precise numerical values to risk parameters (see for instance [30–32]). Possibility Theory was initiated by Zadeh [4], and later developed by several authors among which Dubois and Prade [6]. In its variants, it has found applications in several domains including project investment decision [33], project network analysis [34], contract decision making [35] safety performance [36]. Within risk assessment, Possibility Theory has been used in various hybrid forms and in a limited number of use cases (see for instance [37]). From the methodological point of view, several works have focussed on the problem of merging data and expert opinion [38, 39]. In the context of general decision making with possible applications to risk management, most works focus on the joint propagation of variability and imprecision in the risk assessment process [3, 40, 41].

3 Elements of Possibility Theory

3.1 Possibility Distributions

The basic entity of Possibility Theory is the *possibility distribution*, denoted π, which represents the subjective knowledge of an agent about the actual state of affairs x of a quantity. It consists in a mapping π from a set of states of affairs S to a totally ordered scale of plausibility, such as the unit interval $\pi_x : s \in S \rightarrow \pi(s) \in [0, 1]$ and it distinguishes what is plausible from what is less plausible for the ill-known quantity x according to the following conventions:

- $\pi_x(s) = 0$ means that the state s is rejected as *impossible* for x;
- $\pi_x(s) = 1$ means that the state s is *totally possible* (= unsurprising) for x.

If the state space is complete, at least one of its elements should be the actual state (closed word hypothesis), so that at least one state is totally possible: this is referred to as the *normalisation condition*: $\max_{s \in S} \pi_x(s) = 1$. Distinct values within S may simultaneously have a degree of possibility equal to 1. This framework can represent extreme forms of partial knowledge:

- *complete knowledge*: for some state s_0, $\pi(s_0) = 1$ and for all the others $\pi(s) = 0$, i.e. only s_0 is possible;
- *complete ignorance*: $\pi(s) = 1 \quad \forall s \in S$, i.e. all the states are totally possible.

The simplest non-extreme form of a *possibility distribution* on the set S is the bipolar characteristic function of a subset E of S, i.e. $\pi(x \in E) = 1$ and $\pi(x \notin E) = 0$. It models the situation when all that is known about x is that it cannot lie outside E. This type of possibility distribution is naturally obtained from experts stating that a quantity x lies between values a and b: in which case E is the interval $[a, b]$. However, this binary representation is not entirely satisfactory: sometimes, even the widest set of possible values does not rule out some residual possibility that the value of x lies outside it: so it is natural to use a graded notion of possibility. In this case, *formally, a possibility distribution π coincides with the membership function μ_F of a fuzzy subset $F \subseteq S$, such that $\mu_F(s) = \pi_x(s)$.* The core of a possibility distribution $\pi_x(s)$ is the set of s such that $\pi_x(s) = 1$; the support is the set such that $\pi_x(s) > 0$.

3.2 Possibility and Necessity

Based on the knowledge captured by the *possibility distribution*, Possibility Theory provides two dual evaluations of the likelihood of an event (for instance, that the actual value x of an ill-know quantity, should lie within a certain interval): the *possibility* Π and the *necessity* N of the event. Given a subset of states A and the event "*the actual value x of the unknown quantity lies in A*" the normalized measure of possibility Π and necessity N are defined from the possibility distribution π : $S \rightarrow [0, 1]$ such that $\sup_{s \in S} \pi_x(s) = 1$ as follows:

$$\Pi(A) = \sup_{s \in A} \pi_x(s) \quad N(A) = 1 - \Pi(\bar{A}) = \inf_{s \notin A}(1 - \pi_x(s)) \tag{1}$$

Π tells to what extent *at least one element of A* is consistent with the knowledge π_x (i.e. is possible), while $N(A)$ tells to what extent *no element outside A is possible* (i.e. to what extent A is implied by the knowledge π_x), i.e. the *possibility* measure refers to the idea of *plausibility*, while the dual *necessity* measure refer to the idea of *certainty*. They are dually related: $N(A) = 1 - \Pi(A^c)$, where A^c is the complement of A, the certainty of an event reflects a lack of plausibility of its opposite. However they cannot be obtained one from another. This represents a remarkable difference with Probability Theory, where probability is a self-dual measure (in probability theory $Pr(A) = 1 - Pr(\bar{A})$), however, degrees of necessity can be equated to lower probability bounds and degrees of possibility to upper probability bounds.

The possibility measure and the necessity measure verify respectively the following *maxivity* and the *minitivity* axioms:

$$\Pi(A \cup B) = \max(\Pi(A), \Pi(B)) \quad N(A \cap B) = \min(N(A), N(B)) \quad \forall A, B \subseteq S$$

3.3 Possibility Propagation in Risk Assessment

Most Risk Analysis methodologies prescribe to proceed by a bottom-up compositional approach towards the risk estimate. One should start by breaking the system/process under analysis down to the elementary component level; then for each component, one should list all the possible failure modes; obtain the failure likelihood of each mode and each component; propagate that information from component level to the next level of hierarchy, up to system level so as to obtain eventually, integrating impacts in the calculation, the risk value for each failure mode. At this point, depending on the type of decision to be supported by the analysis, one can take different steps, for instance prioritise risks by failure mode or by component and state the degree of conformance to the objectives or take consequent actions.

In risk assessment the following issues are considered fundamental:

1. Representing the available information faithfully;
2. Accounting for dependencies, correlations between the parameters
3. Choosing a suitable propagation technique.

3.3.1 Input Variables and Their Relationships

Most of the information relevant to the risk assessment of the process of interest come from expert opinion elicitation (for a review of elicitation methods see [42]): for this kind of information the possibilistic representation is commonly deemed more effective than the probabilistic one, since it is difficult for experts to assign precise numerical values to risk parameters. If different opinions about a parameter are provided by different experts or sources, they can be merged by the use of suitable operators or knowledge fusion procedures [38]. In order to illustrate our methodology, we assume, for sake of simplicity, that a single expert has been consulted, and that, provided with the available knowledge, (s)he has expressed its opinion in terms of possibility distributions over the space of the possible events.

Also the assumptions on dependencies can be elicited from expert opinion. The most conservative attitude in absence of contrary expert statements is to make the least specific assumption, corresponding to the hypothesis that the events are separable. For illustrative purposes, hereafter, we adopt this simplifying assumption.

3.3.2 Reliability Propagation

Knowledge propagation, from the system/process components level description to the system/process as a whole, can be carried on through the application of logical aggregators and functional aggregators defined on the basis of the extension principle [43]. For instance, following [44, 45] (other possibilistic approaches to reliability are reviewed in [46]) one can perform a fault-tree based propagation of the *possibility of failure*: from the possibility distributions of a variable describing the failure of individual system/process components it is possible to work out the possibility distribution of variables describing the failure of the whole system/process.

Consider a reliability model where a system component at any instant of time is in one of two states: perfectly functioning $s = 1$ or completely failed $s = 0$, i.e. $s \in S = \{0, 1\} = \{fail, work\}$. Assume event behaviour is fully characterised in terms of possibility measures: this set up is called PosBiSt (from "Possibilistic, Binary States"). Assume that expert opinion has assigned to the ill-known state x_i of the i-th component the possibilities $r_i \equiv \pi(x_i = 1)$ and $u_1 \equiv \pi(x_i = 0)$: here r_i is called *possibilistic reliability* of the component, while u_i is called *possibilistic unreliability* of the component. In a possibilistic setting in general $r_i \neq 1 - u_i$ so one needs to deal with the two quantities separately [47].

If the system state is completely determined by the states of its components, the structure function of an n components system can be written $G = G(\pi_{x_1}, \ldots, \pi_{x_i}, \ldots \pi_{x_n})$. Also for the ill-known system state G we adopt the convention $G = 1$ if it is functioning and $G = 0$ if it is failed and define the system Possibilistic Reliability as $R_G = \pi(G = 1)$ and the system Possibilistic Unreliability as $U_G = \pi(G = 0)$.

For instance, consider the important class of systems whose structure function is composed of only AND and OR gates of the Boolean states of its components. Among them a common class is the class of k-out-of-n systems. A system with n unrelated components is said to have a k-out-of-n G structure if it operates when at least k components operate: fully parallel systems are 1-out-of-n G systems, fully series systems are n-out-of-n G systems. For general k-out-of-n systems the reliability and unreliability can be computed as follows. The Possibilistic Reliability $R_G^{k/n}$ of a k-out-of-n system based on a set A of components whose possibilistic reliabilities form a strictly descending ordered list $(r_1, r_2, \ldots, r_n)$ and whose possibilistic unreliabilities form a strictly ascending ordered list $(u_1, u_2, \ldots, u_n)$ are respectively

$$R_G^{k/n} = \max_{E \in 2^A : |E| = k} \left(\min_{i \in E}(r_i) \right) = r_k \qquad r_i > r_{(i-1)} \; \forall i \in \{2, n\}$$

$$U_G^{k/n} = \max_{E \in 2^A : |E| = (n-k+1)} \left(\min_{i \in E}(u_i) \right) = u_k \qquad u_i < u_{(i-1)} \; \forall i \in \{2, n\}$$

Arbitrarily complex fault trees consisting—whose details depends on the system/process structure—can be used to propagate the uncertainty through any specified system/process, so as to obtain the possibility of system/process failure events.

3.3.3 Risk Assessment

The propagation of possibilistic reliability from the components to the system allows to obtain the possibility $\pi_G(G = 0)$ of the adverse event $(G = 0)$ (failure) at system/process level. However, to compute risk one needs also to assess impacts.

Impact Assessment

The impact of the adverse events can be estimated by different approaches: we will illustrate the approach—based on Value of Information analysis—used by our methodology in Sect. 5.

Also the knowledge about the impact is suitable to a possibilistic representation. Typically the output of impact assessment is a fuzzy possibility distribution π_I over the economical value of the incurred damage. The shape of the distribution is obtained by propagation of uncertainty from the input parameters of the impact to the overall impact through fuzzy algebraic operations.

The propagation problem can be stated as follows. Let the impact $I : \mathbb{R}^n \to \mathbb{R}$ be a function of m variables y_i (i.e. $y = (y_1, \ldots, y_m)$)—possibly represented as an array of fuzzy sets $(\pi_{y_1}, \ldots, \pi_{y_m})$ carrying the semantics of possibility distributions. The problem consists in carrying over to $I(y)$ the uncertainty attached to the variables so as to obtain I's possibility distribution, the distribution $\pi_I(y)$—with the least possible loss of initial information.

Risk Computation

Once the knowledge about the *likelihood* of the adverse event A is represented in terms of the possibility distribution π_G and the knowledge about its *impact* is represented in terms of the possibility distribution π_I a possibility distribution π_R of the risk R can be obtained by the product-convolution of the two distributions, corresponding to the product of the two variables: $\pi_R = \pi_G \otimes \pi_I$.

The output of this operation is a possibility distribution $\pi_R(R = r)$ over the admissible economic values of risk. Based on this distribution one can make further computations to support the decision process. Given $\pi_R(r)$, one can choose different approaches to support the decision. For instance if the aim of the risk analysis is to assess whether the risk exceeds some predefined threshold t, one can define the event F as $r > t$ and compute possibility $\Pi(F)$ and necessity $N(F)$ of the event F. A decision can then be based, for instance, on the synthesis value given by their average or on a different synthesis value. A technique for obtaining synthesis values suitable to risk assessment consists in taking account the subjective risk aversion of the decision maker by adopting the so called Hurwicz criterion, generalised to possibility distributions as proposed by Gyuonnet et al. [48].

The methodology described in this chapter aims to allow comparison among the residual risks of competing risk mitigation strategies. In this case the decision can be taken either based on the ordering of synthesis indicators or based on ranking criteria for fuzzy intervals [49] as proposed by Dubois and Prade (called the *four grades of dominance* method) [50].

4 Process Model

Let us now formalise our notion of business process model. Since the aim of our model is enabling risk assessment, the process representation will focus on risk-related rather than design-related aspects. We start by representing the business process' set of actors as a set $A = \{A_1, \ldots, A_n\}$. Each actor A_j holds a (possibly empty) information item $INFO_j$ whose content is used to generate messages to be exchanged during the business process' execution. Also, we denote by $\{I_{j,k}\}$ the impact of the disclosure of $INFO_j$ to A_k (as assessed by A_j). In principle, this impact can be positive or negative, and can depend on a number of factors, including the content of $INFO_j$ or of other information items.[1] In our view, security controls (when present) are an integral part of the business process definition. In order to be able to represent a complete set of security controls, message exchange in our process model is a general *time-stamped choreography* [51] consisting of:

- *Messages*, i.e. triples (A_i, A_j, m_{ts}), where m_{ts} is (a part of) an *INFO* item and *ts* is an integer representing a discrete time.[2]
- *Local computations* $(A_i, f(), INFO_{i,ts})$ i.e. functions computed by actors on (portions of) locally held information at a given time.

Figure 1 (left) shows an illustrative process model.

---

[1] Of course, the impact of disclosing an empty item is always 0.

[2] For the sake of simplicity, in our model we assume synchronous clocks and instant message delivery.

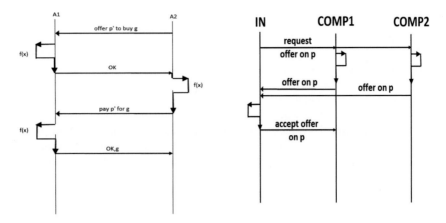

Fig. 1 *Left* an illustrative process model. *Right* a sample cloud process model

4.1 Process Model Assumptions

Our process model is completed by some additional assumptions. Here β denotes the possibility of an event, however assessed (for our own possibility assessment technique see Sect. 5.2):

- *Protocol efficacy*: Given a message delivery (A_s, A_d, m_{ts}), with $m_{ts} = INFO_s$ then $\beta_i(E_{sd}) = MAX$ for all actors A_i.
- *Information completeness*: Given a message delivery (A_s, A_d, m_{ts}), with $m_{ts} \leq INFO_s$, then $\beta_i(E_{sd}) = 0$
- *Strong local computation transparency*: Given a local computation $(A_i, f(), INFO_{i,ts})$, then $INFO_i = INFO_{i,ts} \cup f(INFO_{i,ts} \cup S_f)$ for $t \geq ts$, where S_f is the *specification* of f as an algorithm or a closed formula.
- *Belief propagation*: Given a message delivery (A_s, A_d, m_{ts}), then for $t \geq ts$, $\beta_i(E_{sk}) = \beta_i(C(A_d, A_k))$ for $k \neq d$, where $C(A_d, A_k)$ denotes the event of information sharing between A_d and A_k.

It is important to remark that our *local computation transparency* assumption implies that any actor computing a function $f()$ over its local data becomes aware of the results of that function as well as of its *specification* S_f, represented e.g. as a computer program. However, research has shown that this assumption may be weakened by *obfuscation* or *garbling* techniques [52].

4.2 Garbling Outsourcing Scheme

Garbled circuits, a classical idea rooted in early work by Andrew Yao, are a well-known example of obfuscation techniques. Here, we follow the literature [52] to briefly describe a *garbling outsourcing scheme* corresponding to Yao's garbling technique. The purpose of our simplified description is to show how *obfuscation* is represented within our process model. Let us assume Alice wants Bob to compute for her a function $f()$ on a set of inputs, some of which are held by herself and others by Bob, without sharing with Bob the function specification S_f. At a high level of abstraction, the scheme works as follows: Alice creates a "garbled circuit", i.e. the specification S'_f of a garbled function $f'()$ having the same input-output table as $f()$, and sends it to Bob. Bob uses S'_f to build $f'()$, compute it with his inputs B and returns the result to Alice. The result of $f'(B, x)$ evaluation with $x = A$ (where A is Alice's inputs) coincides with the function $f()$ that Alice wanted Bob to compute; but by computing $f'()$, Bob has learnt nothing about S_f. Note that in this scheme Alice does not send her inputs to Bob; rather, her inputs are encoded into the "garbled circuit" in such a way that Bob cannot determine what they are. As an example, assume that Bob has $x = 2$ bits, (a, b), and Alice has $y = 2$ bits, (c, d). The function $f()$ is:

$$f(x, y) = (a + c) \lor (b + d) \tag{2}$$

For the construction of the garbled circuit, one simply constructs a new truth table for each gate in the original circuit. A sample truth table for an AND gate is shown Fig. 2, with inputs p, q and output z. Alice picks two random keys for each wire and obtains the garbled truth table by encrypting the output-wire key with the corresponding pair of input-wire keys.

After Bob has received the garbled specification $S_{f'}$ and the corresponding truth tables, he still needs Alice's inputs before he can evaluate the function. Bob can get these inputs using a 1-out-of-2 instantiation of Rabin's *oblivious transfer* protocol.[3] Once Bob has received the input values from Alice via the oblivious transfer protocol, he can "decrypt" each of the gates, and using his own inputs he can evaluate the circuit. Today, efficient garbling schemes are available achieving privacy as well as obliviousness and authenticity, the latter properties being needed for private and verifiable outsourcing of computation. Highly efficient block-cipher-based instantiations of garbling schemes have been described in the literature. For our purposes, it is sufficient to observe that when a garbling outsourcing scheme is in force within a process, a weaker assumption (*weak local computation transparency*) can be adopted for our business process model, where the party executing a local computation $f()$ learns the output of the function, but not its specification. More formally, let P be a process including a local computation $(A_i, f(), INFO_{i,ts})$. Let $G()$ be a functional acting on the $f()$ function specification S_f, so that $G(S_f) = S_{f'}$.

---

[3]In an oblivious transfer protocol, a sender transfers one of potentially many pieces of information to a receiver, but remains oblivious as to what piece (if any) has actually been transferred [53].

input1	input2	output	garbled computation
k_q^0	k_p^0	k_z^0	$E_{k_q^0}(E_{k_p^0}(k_z^0))$
k_q^0	k_p^1	k_z^0	$E_{k_q^0}(E_{k_p^1}(k_z^0))$
k_q^1	k_p^0	k_z^0	$E_{k_q^1}(E_{k_p^0}(k_z^0))$
k_q^1	k_p^1	k_z^1	$E_{k_q^1}(E_{k_p^1}(k_z^1))$

Fig. 2 Garbled computation for an AND gate

We call $G(S_f) = S_{f'}$ a *garbled specification* of $f()$ if and only if $f'(x) = f(x)$ for all inputs x and:

$$t \geq ts : (A_i, f(), INFO_{i,ts}) \rightarrow INFO_i = INFO_{i,ts} \cup f'(INFO_{i,ts})$$

It is important to remark that the computation of $G(S_f) = S_f'$ can itself be a local *computation of the process P*. This way, any actor can outsource a local computation to another actor, who will compute the garbled function under our weak transparency assumption.

5 Methodology for Disclosure-Risk Assessment in Cloud Processes

In this Section we describe the key elements of our quantitative risk assessment approach, namely, the identification of the adverse events and threats, the estimation of threats' possibility distributions and of their impacts.

5.1 The Threat Space

Any risk model must clearly specify the event space where risk will be quantified. In this chapter we focus on a single, albeit large, family of threats, namely *data process-related leakage threats*, i.e. *the disclosure of one or more information items to be exchanged in a multi-party protocol to participating parties who are not the originally intended recipients*. A major adverse event happens when actors (including service and cloud providers) put together the partial information they hold to reconstruct knowledge that is not available to them when taken individually. We remark that this adverse event is not caused uniquely by collusion among rogue participants. Indeed, different parties may put together their information for other reasons, including:

- eDisclosure, i.e. the mandatory process of disclosing information to adversaries during litigation[4]

---

[4]http://www.edisclosureinformation.co.uk.

- An information request from a regulatory authority[5]
- Inadvertent or dysfunctional behaviour of employees.

For the first factor, data sharing imposed by courts of law may generate leaks that are difficult to identify a priori even for experienced security auditors. The second factor—the intervention of a regulatory authority—is also difficult to predict. For instance, e-mails containing bids for a auction held in one country may be stored on a server located in another jurisdiction, where a regulatory authority can ask the service provider—for reasons unrelated to the auction—full access to the storage of the mail server, without informing the auctioneer. This way, a third party would get to know in advance the outcome of the auction.

As one would expect, the third factor has the strongest documentary evidence. A global security study on data leakage, commissioned by Cisco and conducted by a U.S.-based market research firm [54] polled more than 2000 employees and information technology professionals in 10 countries, including major EU markets. The study identified the adverse event of unwanted information sharing, related to sloppy implementations of interchange protocols, or intentional communication with unauthorized parties. For instance, a plain-text email containing a business offer sent in good faith through a "secure" cloud-based mail service poses a danger if disclosed by the cloud provider to a competitor of the original sender. Today, it is very challenging even for experienced process owners to fully identify, analyze and handle data leakage risks, due to the complexity and diversity of business processes and of the underlying IT systems; the trend toward outsourcing and the cloud is further blurring the scenario. Many organizations have little visibility into where their confidential data is stored on the cloud or control over where that data is transferred during the execution of a process. Even when insight is available, organizations often lack a clear methodology to assess whether the process involves an acceptable level of risk. The methodology and models presented in this chapter are aimed at filling this gap.

5.2 Possibilistic Model of Information Disclosure

In this Section we will discuss how to estimate the possibility that subsets of process actors *collude*, i.e. put together the information they know. We make use of the notion that the dysfunctional behaviour of actors in a business process is often due the *unfairness of the redistribution of payoffs* in the process, (e.g. a benefit allocation structure that responds to organisation efficiency more than to fairness). Our technique does not rely on frequency-based probability estimates, rather we derive the possibility based on the degree ϕ of unfairness in the process resource allocation and then associate to each subset of actors the possibility distribution of

---

[5]Whether data is on premises or in the cloud, the obligation to comply with the demands of the court or regulatory authorities remains essentially the same.

the event in which the subset members decide to put together the information they hold.

The problem of how profits of a coalition should be redistributed is a well-known one. A principled solution can be obtained by attributing to each actor an amount corresponding to the actor's *Shapley Value* [55]. Indeed, given a coalition, the contributions of the actors to the process, and the value of the surplus value produced by the process, the Shapley value yields a unique ideal allocation of that value fulfilling some largely accepted requirements. With N actors, this solution can be visualized as a point on the hyperplane of the feasible allocations in an N-dimensional space.

This approach can be applied to the actors of an organization to find the fairness point and compare it to the point representing the current allocation of the value in the organization: the distance between these two points provides for each actor an estimate of its individual dissatisfaction. The farther the two points, the more likely there will be dysfunctional behavior on the part of that actor. Elicitation of expert opinion can transform this information, into a possibility distribution for the defection of the individual actor (for a use of a similar method within a probabilistic approach see [56–62]).

Hereafter first we provide a formal definition of the Shapley Value and of the computation of the dissatisfaction parameter, then illustrate the use of this information in the determination of the actors' unreliability.

5.2.1 The Shapley Value

The Shapley Value, is an allocation respecting some generally accepted criterion of fairness. The idea behind it is that each party taking part to a process should be given a payoff equal to the average of the contribution that she could make considering each of the possible coalitions underlying the process.

In Game Theory it is customary to call *security level* of a coalition C the quantity $l(C)$ expressing the total surplus that its members can achieve on their own even if the non-members took the action that was the worst from C's perspective.

Let us consider a general game with a set $\mathcal{N}$ of N participants. The Shapley Value is defined as an allocation of payoffs: a payoff v_i for each actor $i \in \mathcal{N}$. Any subset of players in $\mathcal{N}$ is a potential *coalition* C. A coalition can strike deals among its own members to exploit all the available knowledge for mutual advantage. Combinatorially, there are $(2^N - 1)$ possible coalitions altogether. In order to produce each coalition one has to run ideally over all the permutations of actors: each ideal ordering of the actors corresponds to a non-decreasing surplus value achieved by the members. One has to take note of the added value introduced by the actor i, whose Shapley Value is being computed, i.e. one has to compute the added value $\Delta_i(C) \equiv (l(\{C \cup i\}) - l(C))$ which is defined by the difference in security levels. The Shapley Value for the actor is then computed as the average over all permutations of its added values:

$$v_i^{Shapley} = \frac{1}{N!} \sum_{\sigma} \Delta_i(\sigma) \qquad (3)$$

where the index σ runs over all the permutations of N objects.

Now let $v_i^{factual}$ be the actual resource allocation for an actor. If the difference between this quantity and the actor's Shapley Value is positive, the actor is under-rewarded for her contribution: this situation may feed its propensity toward a defection; if, instead, this difference is negative, the actor is over-rewarded and the discrepancy will not contribute to its propensity towards defection. One needs also to relate the discrepancy, to the absolute value of $v_i^{Shapley}$. For all the above considerations, the dissatisfaction parameter ϕ_i for an actor i can be defined as follows:

$$\phi_i \equiv \theta\left(v_i^{Shapley} - v_i^{factual}\right)/v_i^{Shapley} \qquad (4)$$

where $\theta(z)$ is such that $\theta(z) \equiv 0$ if $z < 0$ and as $\theta(z) \equiv z$ otherwise.

Elicitation of Expert Opinions

Following the conventions used so far, we adopt the following notation: x_i indicates the ill known quantity representing the state of the individual actor, the actor can be *defecting*, i.e. contribute to data disclosures ($x_i = 0$) or *non-defecting*, i.e. behaving correctly with respect to data disclosures ($x_i = 1$); the possibility distribution for the individual actor, given its ϕ is denoted $\pi_{x_i}(\cdot|\phi_i) : s \in \{0, 1\} \rightarrow \pi_{x_i}(s|\phi_i) \in [0, 1]$. Based on this one can define, in correspondence to the two values of s, two distinct functions of ϕ: the *possibilistic reliability* function as a function of ϕ, i.e. $r_i(\phi) \equiv \pi_{x_i}(s = 1|\phi)$, and the possibilistic The above defined dissatisfaction parameter ϕ can be translated into possibilistic reliability $r(\phi)$ and unreliability $u(\phi)$ based on the elicitation of expert opinion and the application of obvious constraints—namely $r(\phi = 0) = 1$ and $u(\phi = 0) = 0$, the non-increasing character of $r(\phi)$ and non-decreasing character of $u(\phi)$.

We note, in passing, that due to the monotonicity of r and u, already the value of the parameter ϕ alone, without further elicitation of knowledge, can be used to rank the individual actors' reliabilities and unreliabilities. However expressing, in addition, expert knowledge in terms of r and u enables the quantitative assessment of the disclosure risk.

The elicitation of experts' knowledge in order to obtain a function of a continuous variable is a well-known problem. In our methodology we use a technique based on the *Bézier curves*. Those curves have often been used in in the elicitation of expert opinion (see for instance [63, 64]) also in the form of membership functions [65]: they can be used to approximate a smooth (continuously differentiable) function on a bounded interval up to an arbitrary level of detail by forcing the curve to pass in the vicinity of an arbitrarily high number of control points (in two-dimensional Euclidean space) selected by an expert. Overall, the computation of the possibility of defection for the individual actor involves the following steps:

- Expert opinion is elicited to determine the possibilistic reliability $r(\phi)$ and unreliability $u(\phi)$ as a function of the dissatisfaction parameter ϕ: Bézier curve based methods are suitable candidates for eliciting expert opinion.
- Given a specific instance of the collaborative process definition, and the actual allocation to the actors of the resources deriving from the process surplus, the numerical value of ϕ_i is computed for each actor.
- Based on $r(\phi)$ and $u(\phi)$ we compute possibilistic reliability r_i and unreliability u_i for each actor.

Note that the individual level unreliabilities already hint at the weak points of the process: when units of valuable information are fully known by individual process actors, the individual unreliabilities can be used to quantitatively assess the risk of its disclosure. Furthermore, this knowledge can be used to support the many-actor level risk assessment. As it has been shown in Sect. 3.3.2, from individual level (un) reliabilities one can obtain many-actors (un)reliabilities based on the structure function of the failure model. The main results can be translated into the possibility of disclosure. For instance given the reliabilities r_i and unreliabilities u_i of n actors

- (*series* analogy) if the failure model is such that it is necessary that all the actors not to defect in order for the information to be disclosed, i.e. if it is sufficient that $k = 1$ actor defects for the disclosure to take place, then the possibilistic reliability is $r = \min_i(r_i) = r_n$ and the possibilistic unreliability is $\max_i(u_i) = u_1$
- (*parallel* analogy) if the failure model is such that it is *sufficient* that $k = 1$ actor does not defect for the disclosure not to take place, i.e. it is *necessary that all* the actors defect in order for the information to be disclosed, then the possibilistic reliability is $r = \max_i(r_i) = r_1$ and the possibilistic unreliability is $\min_i(u_i) = u_n$
- (*k*-out-of-*n* analogy): in this case the failure model is such that it is sufficient that k (out of n) actors do not defect for the disclosure not to take place, i.e. it is necessary that $(n - k + 1)$ actors defect in order for the information to be disclosed; as a consequence, the possibilistic reliability is equal to the k-th largest reliability r_k and the possibilistic unreliability is equal to the k-th smallest unreliability u_k.

5.3 Impact Assessment by Value of Information Analysis

The technique we use for estimating impact of information disclosure relies on quantifying the *Value of Information* (VoI) for each knowledge item (or set of items) potentially reconstructed by a subset of colluding process actors. Also the estimated value of a knowledge item has an intrinsic possibilistic character since it will be known to lie in a range but its precise economical value will depend from several incompletely known factors.

5.3.1 Value of Information Analysis

VoI has been defined as the *analytic framework used to establish the value of acquiring additional information to solve a decision problem*. In the risk management domain, VoI has been successfully used since the Sixties in several areas of research including engineering and environmental risk analysis [66]. From a purely rational perspective, it is clear that acquiring extra information is only useful for an actor A if knowing it can significantly modify its behaviour.

Classic VoI analysis typically involves constructing a complex decision-analytic model to fully characterise all information items available to each process actor, the loss each actor would incur should these items become known to other actors, the costs of interventions that could be executed to prevent them. This comprehensive approach to VoI often turns out to be prohibitively expensive for use in prioritising interventions [67]. As alternatives to full VoI, we identified three approaches to analysing the value of information that are less burdensome:

1. The *conceptual* approach to VoI, where context information is used to provide informative bounds on the value of information without formally quantifying it through modeling. For instance, the VoI of the design information about a device that is already available on the market cannot be higher than the cost of reverse-engineering the device itself;
2. The *minimal* approach to VoI, which is possible when evidence of the net benefit of holding a piece of information, are readily available from existing research. For example, the VoI of the design information about a device that is currently available on the market cannot be higher than the net profit coming from its sales to its current supplier.
3. The *maximal* modeling approach to VoI, where the value of an information item is estimated from previous VoI studies concerning similar information in different contexts. For instance, the VoI of the design information about a solid-state storage device is quantified according to previous VoI studies on disks.

These three low-cost VoI methods can be readily applied in priority-setting of risk-mitigation countermeasure, and raises the question about how the use of VoI to assess disclosure risk in the framework of our methodology.

Value of Total Information and Value of Partial Information

Here, we take a process-oriented view of VoI, in order to assess the impact of information disclosure. Let us consider once again a set of actors $A = \{A_1, \ldots, A_n\}$ who take part to a business process P, and the expected benefit for each actor A_k, Ben_{A_k} resulting from the execution of P. The starting point of our VoI analysis of

P is to consider the *Value of Total Information* (VoTI), i.e. answering the question "What would be the change to Ben_{A_i} should A_i know all information (local memory plus messages) held by the other actors of *P*?". If there is no such change, then achieving extra information is worthless. If such a change exists, then the impact on A_k of A_i's $(i \neq k)$ complete knowledge can be estimated as the corresponding change in the value of Ben_{A_k}.

For the security-aware process designer, our simple VoTI provides a useful upper bound, because it tells the maximum value that any information held by other actors may have for each participant to *P*. If that value is negligible, or achieving that information would cost more than that, a rational actor will not pursue disclosure any further (i.e., it would not enter agreements for information sharing with other actors).

A different type of check involves looking at the *Value of Partial Information* (VoPI). For any process participant A_i, getting to know some information beyond the one that is strictly necessary to carry out its part in the process (e.g., the messages exchanged among other actors, or the content of another actor's local memory) may or may not bring a benefit, i.e. a change in Ben_{A_i}. For each subset *K* of knowledge items used in the process, VoPI focuses on (i) checking whether the benefit of knowing *K* would match the cost of collecting it and (ii) quantifying the impact of each actor A_i getting to know *K* on the benefits Ben_{A_k} of the other participants (for $i \neq k$).

5.3.2 Possibilistic Value of Information

The fact that the output of the Value of Information analysis is in nature possibilistic and fuzzy has been broadly recognised and accepted in several domains [68–72]. Indeed, the knowledge of an expert who has performed VoI analysis can be best conveyed by means of a possibility distribution (in this case it will not be a distribution over discrete values but over continuous values). In line with the traditional representation of the risk as the product of the likelihood of the adverse event by the impact of the event, one can represent the possibilistic risk by the (fuzzy) product of the possibility distribution of the event and the possibility distribution of the values that the impact can take. This point is discussed further in the next Section.

5.4 The Overall Methodology

In our approach, managing risks related to the execution of a business process *P* in presence of threats constitutes itself a process (usually called *risk management process*, in symbols $M_{R(P)}$) where alternative techniques for dealing with threats are compared according to a procedure. The output of $M_{R(P)}$ is a *risk mitigation strategy*, which consists of modifications to P that have some effect on the risk of executing it, including the introduction or removal of security controls.

In this Section, we describe our methodology for comparing alternative risk mitigation strategies. This methodology does not provide specific guidance on the choice of mechanisms that will actually counter the threats; rather, it allows comparing the residual risk of competing risk strategies. Although qualitative comparison is supported, the methodology aims to quantitative cost-benefit calculations, assessments of risk tolerance, and quantification of preferences involved in $M_{R(P)}$. We are now ready to provide a step-by-step description:

- The first step is the *stakeholder identification*, where we identify the actor set A of our business process P and compute its power set 2^A. In our approach, process stakeholders include all participants to P. Namely, our actor set includes *all actors who, according to the risk assessor, may in any way get the capability of reading (or writing) information shared during P's execution.* As we shall see in Sect. 6.1, actors in A can be further refined by type according to their role in the computation.
- The second step consists in the *formalization of the business process model*, using the syntax introduced in Sect. 4, which represents two types of actions: (i) *message exchanges* and (ii) *local computations*. It is important to remark that while execution-oriented process models usually contain control structures like conditions and loops [73], our process model syntax expresses all possible execution paths *independently*, i.e. as separate models. The next step takes care of this.
- The third step consists of *process streamlining*, which includes *loop unrolling* and *re-encoding of conditions as parallel paths*. Here we do not enter into the details of business process streamlining, as process improvement techniques have been deeply studied since the Eighties and are discussed in detail in the technical literature (see for instance the rich bibliography of [73]). However, software toolkits supporting our methodology will have to provide guidance w.r. t. process streamlining.
- The fourth step, *identifying reconstructible knowledge*, consists in computing the *knowledge set* K_S for each subset $S \in 2^A$. Each knowledge set includes *all the knowledge that members of A can achieve by putting together the information they hold.*
- The fifth step consists in *estimating the disclosure impact* of K_S for each subset S in 2^A at each step of the business process P, and to give it a possibilistic representation (see Sect. 5.3)
- The sixth step consists in *estimating the possibility distribution for the defection* for each individual actor, based on process related information, on elicitation of expert opinion and on the techniques shown in Sect. 5.2. Also to this information will be given a possibilistic representation.
- The seventh step consists in *estimating the collusion possibility distribution* for each subset S in 2^A at each step of the process P (see Sect. 5.1). Once again it is important to remark that this estimate needs to be process-specific (as it will take

into account the micro-economics and social relations underlying P) and take into account multiple causes of collusion, including dysfunctional behavior, intervention of regulatory authority and others. The estimation will be based on on uncertainty propagation methods as those shown in Sect. 3.3.2.

- The eight step consists in *aggregating the products* between (i) the possibility distributions of collusion for each subset S in 2^A and (ii) the possibility distributions of the impact of K_S at each step of the process P, obtaining the *total risk* related to the process, in the form of a possibility distribution, based on which one can take the risk management decisions (see Sect. 3.3.3).

6   A Cloud-Based Case Study

Here we specialize the process model presented in Sect. 4 to describe cloud-based computations. We rely on a variation of Bogdanov et al.'s representation of cloud actors [74].

6.1   The Cloud Process Model

In order to make our representation of our multi-party business process actors suitable for describing cloud-based computations, our actor set A becomes a (non-necessarily disjoint) triple $\{IN, COMP, RES\}$ where IN denotes actors holding non-empty information items (a.k.a. input nodes), while $COMP$ and RES are auxiliary sets of actors (a.k.a. *compute* and *result* actors) whose information items are initially empty. Such actors respectively perform local computations ($COMP$) and publish results (RES). The following constraints—looser versions of the ones in [74]—are in place for our cloud model:

- *Separation of duties*: Sender actors belong to IN and $COMP$ only.
- *Local information integrity*: Any actor can send part of an $INFO$ item it holds entirely, or relay parts it has previously received from other actors.

Figure 1 (right) shows a sample visual representation of a cloud-based process, where a buyer sends messages to two sellers who respond with their offers.

6.2   Possibilistic Assessment of Likelihood and Impact

For each subset $S \in 2^A$ of the set of actors we can now compute the risk of disclosure for information shared within S, at each time t as a product of the impact and the likelihood of the collusion event $E_S(t)$, expressed in terms of the event possibility.

With regard to the former, we will indicate by $I_S(A_k, t)$ the damage that members of S can do the actor $A_k \in A$, using all messages in the process incoming then with timing lower or equal to t. We will indicate by $K_S(t)$ the (possibly empty) *common knowledge* of S, which consists then in the *INFO* items that the actors can jointly reconstruct from the shares they hold at time t. The knowledge $K_S(t)$ can be estimated based on the choreography (and the security features) of the process, the impact can be estimated from expert opinion and expressed in terms of a possibility distribution $\pi_{I_S(A_k, t)}$ over the feasible impact values (a fuzzy set). Hereafter, for sake of notational simplicity, we indicate such possibility distribution by $I_S(A_k, t)$. Furthermore, in the following algebraic expressions, if the individual addends or factors are fuzzy sets, the arithmetic operations have to be understood as a fuzzy operations.

Computing the risk posed by S to A_k also requires estimating the possibility $\pi(E_S(t))$ of members of S having colluded at time t. We will denote this quantity simply by $P_S(t)$. Multiplying the two quantities will yield the possibility distribution of the risk R posed at time t to actor A_k by the collusion of the set S: we will denote this possibility distribution by the shorthand $R(A_k, E_S) = P_S(t)I_S(A_k, t)$.

6.3 Sample Assessments

Let us start with the simplest example: a business process where a client uses a cloud-based computation service to add two integer numbers and another one to publish the result. In this case, we have the actor set $A = (IN1, COMP1, RES)$ where actor $IN1$ holds the information item $INFO_1$ containing the two summands $INFO_1[1]$ and $INFO_1[2]$, actor $COMP1$ is the outsourced service that computes the addition, while actor RES publishes the result. The process is represented by the choreography shown in Fig. 3 (left), where the input actor $IN1$ sends $INFO_1$ to

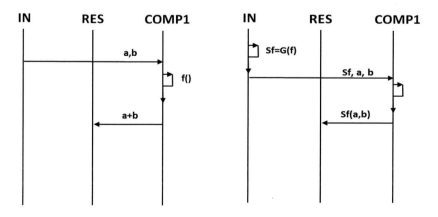

Fig. 3 *Left* sample business process. *Right* choreography for the garbled sum computation

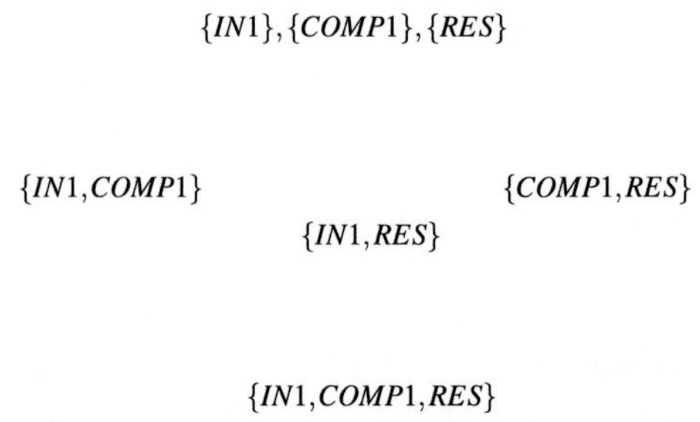

Fig. 4 The Boolean lattice $\{IN1\},\{COMP1\},\{RES\}$
for the considered actor set

$\{IN1,COMP1\}$ $\{COMP1,RES\}$
                          $\{IN1,RES\}$

$\{IN1,COMP1,RES\}$

$COMP1$, who computes the local function $f(INFO_1) = INFO_1[1] + INFO_1[2]$, i.e.
adds the summands, and sends the outcome to the result node RES who outputs it.

According to the definitions given in the previous Section, the (possibly empty)
common knowledge of any subset of actors $S \in 2^A$ at time t, namely $K_S(t)$, is
composed of the $INFO$ items that have been received in their entirety by all
members of S at or before time t. The power set 2^A of the actor set is the simple
Boolean lattice (Fig. 4):

Running our sample business process, we obtain the following knowledge sets
$K_S(t)$ for $t = 1,2,3$ (we omit the formal step $K_\emptyset(0) = K_\emptyset(1) = K_\emptyset(2) = \emptyset$):

The Process Initialization $t = 0$

First Step $t = 1$
Second Step $t = 2$

The disclosure risk estimated by actor $IN1$ at $t = 0$ is zero, as there are no
subsets $X \in 2^A$ such that $K_X(0) \neq \emptyset$ but the singleton $\{IN1\}$, whose only member
coincides with the risk evaluating actor $IN1$ (Figs. 5, 6 and 7).

At $t = 1$, however, there is another singleton such that $K_X(1) \neq \emptyset$, namely the
subset $X = \{COMP1\}$. All the other subsets for which $K_X(1) \neq \emptyset$ can be obtained
by computing the ideal generated by $\{IN1, COMP1\}$ w.r.t. the Boolean lattice's
join ($\cup$), so their contribution to the risk estimation is zero (all their members
separately had the same knowledge as they have when taken together).

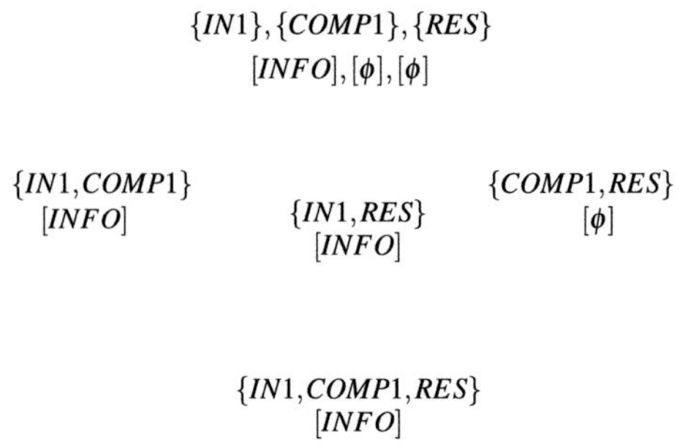

Fig. 5 The knowledge sets at $\{IN1\},\{COMP1\},\{RES\}$
time $t = 0$ $[INFO],[\phi],[\phi]$

$\{IN1,COMP1\}$ $\{COMP1,RES\}$
$[INFO]$ $\{IN1,RES\}$ $[\phi]$
                             $[INFO]$

$\{IN1,COMP1,RES\}$
$[INFO]$

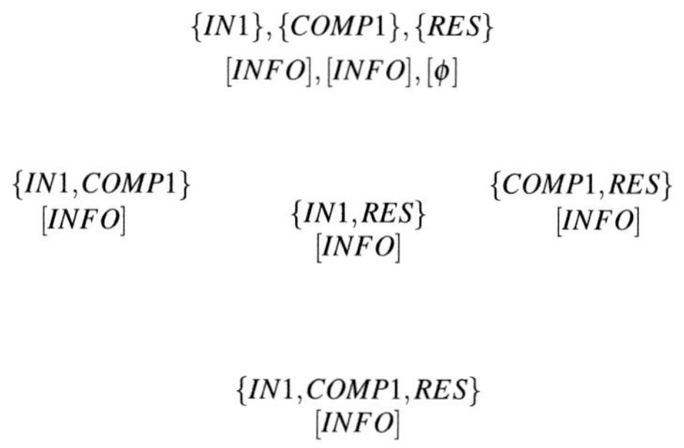

Fig. 6 The knowledge sets at
time $t = 1$

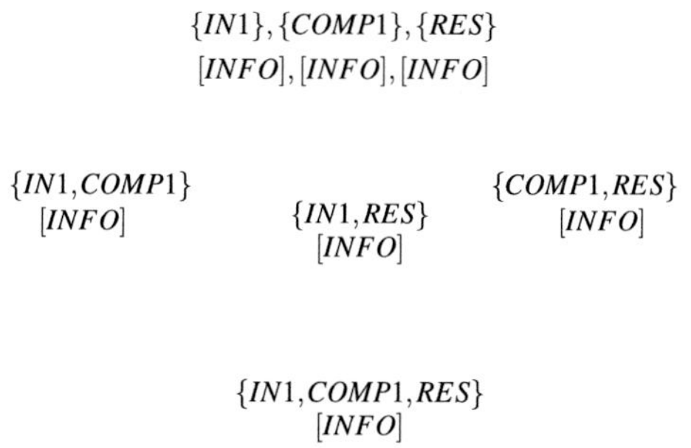

$$\{IN1\}, \{COMP1\}, \{RES\}$$
$$[INFO], [INFO], [\phi]$$

$$\{IN1, COMP1\}$$
$$[INFO]$$

$$\{IN1, RES\}$$
$$[INFO]$$

$$\{COMP1, RES\}$$
$$[INFO]$$

$$\{IN1, COMP1, RES\}$$
$$[INFO]$$

Fig. 7 The knowledge sets at
time $t = 2$

$$\{IN1\}, \{COMP1\}, \{RES\}$$
$$[INFO], [INFO], [INFO]$$

$$\{IN1, COMP1\}$$
$$[INFO]$$

$$\{IN1, RES\}$$
$$[INFO]$$

$$\{COMP1, RES\}$$
$$[INFO]$$

$$\{IN1, COMP1, RES\}$$
$$[INFO]$$

The estimate by actor $IN1$ of disclosure risk in the part of $COMP1$ of information $INFO_1 \cup S_f$ at $t = 1$ can therefore be written as follows:

$$R(A_k, E_S) = R(IN1, E_{\{COMP1\}}) = P_{\{COMP1\}}(1) I_{\{COMP1\}}(IN1, 1)$$

where $P_{\{COMP1\}}(1)$ is the possibility distribution (assessed by $IN1$) that $COMP1$ will disclose at $t = 1$ the information it now holds, i.e. the data $INFO_1$ and the specification S_f of the local function $f()$ it computes (i.e. the addition). $I_{\{COMP1\}}(IN1, 1)$ is the resulting total damage to $IN1$ of the service provider $COMP1$ disclosing what it knows, i.e. the summands $INFO_1$ and the specification S_f. At $t = 2$, another singleton set such that $K_X(1) \neq \emptyset$ pops up, namely $X = \{RES\}$.

Again, all the other subsets for which $K_X(1) \neq \emptyset$ can be obtained by computing the ideal generated by $\{IN1, COMP1, RES\}$ w.r.t. the Boolean lattice's join ($\cup$) (in this case, the entire lattice) so their contribution to the risk estimation is zero (all their members had the same knowledge separately than they have together).

Total risk estimate at $t = 2$ by $IN1$ is therefore:

$$R(A_k, E_S) = R(IN1, E_{\{COMP1\}} \cup E_{\{RES\}}) \tag{5}$$

that becomes:

$$R(A_k, E_S) = P_{\{COMP1\}}(1)I_{\{COMP1\}}(IN1, 1) + P_{\{RES\}}(2)I_{\{RES\}}(IN1, 2)$$

Of course, risk seen by other actors of P can also be evaluated by the same procedure. For instance, the risk estimated by RES at $t = 0$ is related to the singleton subset $\{IN1\}$ (the only one whose knowledge set is not empty; note that in this case, unlike before, it does not coincide with the risk assessor). We get:

$$R(A_k, E_S) = R(RES, E_{\{IN1\}}) = P_{\{IN1\}}(0)I_{\{IN1\}}(RES, 0)$$

In the same line, risk estimated by RES at $t = 1$ can be written as follows:

$$R(A_k, E_S) = R(RES, E_{\{IN1\}} \cup E_{\{COMP1\}})$$

that becomes:

$$R(A_k, E_S) = P_{\{IN1\}}(0)I_{\{IN1\}}(RES, 0) + P_{\{COMP1\}}(1)I_{\{COMP1\}}(RES, 1)$$

6.4 Mitigating Disclosure Risk

In order to mitigate disclosure risk, we apply our *risk management methodology* $M_{R(P)}$ to compare alternative strategies for dealing with risks connected to disclosure threats. The output of $M_{R(P)}$ is a *risk mitigation strategy*, which consists of modifications to the process P (including the deployment of security controls) that have the desired effect on the risk of executing it. While our methodology does not include specific guidance in the choice of such controls, we remark that the user can identify possible changes to the process by searching pattern libraries offering alternative mechanisms for achieving and certifying the security properties of business process and services (e.g. see [75]).[6]

We consider first a pattern of obfuscation of the local function $f()$. Instead of pushing the plain-text specification of S_f to the service provider $COMP1$, actor $IN1$ can use an obfuscation technique for computing the sum. While the obfuscation techniques themselves are outside the scope of this chapter, we remark that a variety of obfuscation mechanisms have been proposed in the literature, including homomorphic encryption, evaluation of branching programs, and Garbled Circuits (GC). GC evaluation and homomorphic encryption are in principle both suitable for obfuscation of simple arithmetics operations like the one in our example.

---

[6]Also, links between security properties and the corresponding threat spaces have been defined in the framework of several certification schemes [76].

Let us assume that a GC technique is used for obfuscating addition (the size of the garbled adder circuit is small, linear in the size of the inputs), and its secure evaluation is efficient, as it is linear in the number of Oblivious Transfers (OT) and in the number of evaluations of a cryptographic hash function (for example SHA-256).[7] In other words, in this representation of our business process P, our functional $G(f)$ denotes a Garbling Outsourcing Scheme mechanism that locally (i.e., within its own trusted environment) computes a garbled function $G(f) = f'()$ corresponding to the sum, then pushes the garbled specification of $f'()$, namely $S_{f'}$ to $COMP1$. The corresponding sample choreography is depicted in Fig. 3 (right).

The subset analysis carried out in Sect. 6.3 is now repeated after applying our modifications to the process P, but we can now apply the weaker version of our computation transparency assumption (Sect. 4). This way, the information disclosed to $COMP1$ does not include the local function specification any more. We get:

$$R(A_k, E_S) = R(IN1, E_{\{COMP1\}}) = P_{\{COMP1\}}(1)I'_{\{COMP1\}}(IN1, 1)$$

where, since the knowledge reconstructible by $COMP1$ is now smaller than before, $I'_{\{COMP1\}} \leq I_{\{COMP1\}}$. The modification to P has therefore decreased risk; however, the amount of such decrease needs to be compared with the combined costs of (i) the local computation of garbling $G(f)$ on the part of $IN1$ and (ii) the additional complexity of computing the garbled function $f'()$—instead of the original addition $f()$—on the part of $COMP1$.

Another version of P that can be envisioned in order to decrease disclosure risk features *multiple service provisioning*, where the confidentiality of $INFO_1$ is increased by outsourcing the computation of $f()$ to multiple services, each getting to know only a portion (a share) of $INFO_1$. In this case of course we will need to extend our actor set to become $A = (IN1, COMP1, COMP2, RES)$ where, once more, at $t = 0$ actor $IN1$ holds the entire information item $INFO_1$ containing the two summands $INFO_1[1]$ and $INFO_1[2]$. The power set 2^A of the actors is (Fig. 8):

In words, the alternative version of our process P can be described as follows:

1. The input actor $IN1$ computes a local function to divide each summand into two shares, obtaining $INFO_1[1, 1]$, $INFO_1[1, 2]$, $INFO_1[2, 1]$, $INFO_1[2, 2]$.
2. $IN1$ sends $INFO_1[1, 1]$ and $INFO_1[2, 2]$ to $COMP1$, and $INFO_1[1, 2]$ and $INFO_1[2, 1]$ to $COMP2$
3. The two computation nodes compute a local function each on the shares they received, $f_{COMP1} = INFO_1[1, 1] + INFO_1[2, 2]$ and $f_{COMP2} = INFO_1[1, 2] + INFO_1[2, 1]$

---

[7]Garbled integer arithmetics has attracted much attention in the past few years [77, 78] both following Yao's original formulation and the alternative Goldreich-Micali-Wigderson (GMW) protocol. Also, [79] summarises several depth-optimised circuit constructions for various standard arithmetic tasks.

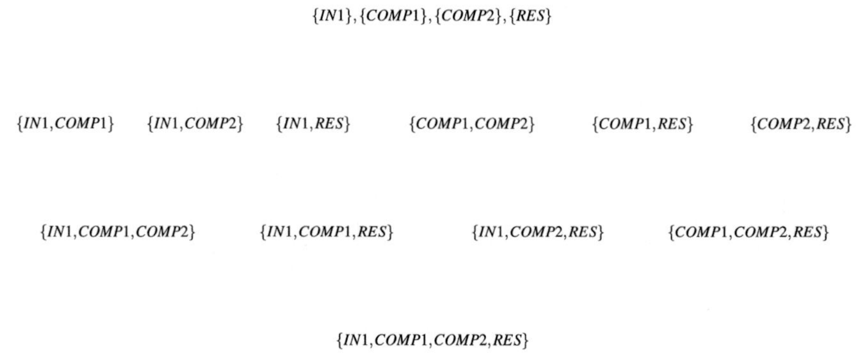

$$\{IN1\}, \{COMP1\}, \{COMP2\}, \{RES\}$$

$\{IN1, COMP1\}$ $\{IN1, COMP2\}$ $\{IN1, RES\}$ $\{COMP1, COMP2\}$ $\{COMP1, RES\}$ $\{COMP2, RES\}$

$\{IN1, COMP1, COMP2\}$ $\{IN1, COMP1, RES\}$ $\{IN1, COMP2, RES\}$ $\{COMP1, COMP2, RES\}$

$$\{IN1, COMP1, COMP2, RES\}$$

Fig. 8 The Boolean lattice for the new actor set

4. The two computation nodes send the results to the result node *RES*
5. *RES* computes $f_{RES} = f_{COMP1} + f_{COMP2}$ and outputs the result

For the sake of simplicity, let us assume that *IN1* will generate two shares of $INFO_1$ using a naive technique, i.e. by taking respectively the Most Significant and the Least Significant Part (MSP-LSP) from the original value $INFO_1$.

For instance, if $INFO_1[1] = 28$ and $INFO_1[2] = 61$, then $COMP_1$ receives $INFO_1[1, 1] = 20$ and $INFO_1[2, 2] = 01$ and computes 21, while $COMP_2$ receives $INFO_1[1, 2] = 08$ and $INFO_1[2, 1] = 60$ and computes 68. Finally, *RES* receives 21 and 68 and computes the final result 89.

Of course, this simplified share generation would not really prevent *COMP* nodes from guessing the original values, so our assumption of *Information completeness*: "Given a message delivery (A_s, A_d, m_{ts}), with $m_{ts} \leq INFO_s$, then $\beta_i(E_{sd}) = 0$" (see Sect. 4) should now be revised here to, say, $\beta_i(E_{sd}) = \frac{1}{10}$, assuming they both *COMP* nodes know that the summands are two-figure integers. However, we will not deal with autonomous guessing in this example, as the threat space we are considering involves only collusions among multiple parties. After defining this revised version of process *P*, we can estimate the knowledge sets corresponding to this new, secured version of the business process. Figures 9, 10, 11, 12 show the evolution of the knowledge sets starting from the initialisation time $t = 0$ to time $t = 3$.

Third Step $t = 3$

Our risk estimate at $t = 1$ by *IN1* is therefore:

$$R(A_k, E_S) = R(IN1, E_{\{COMP1\}} \cup E_{\{COMP1\}}) = P_{\{COMP1\}}(1)I_{\{COMP1\}}(IN1, 1)$$

where, again, $P_{\{COMP1\}}(1)$ is *IN1*'s estimated possibility distribution that *COMP1* will disclose at $t = 1$ the information it now holds, and $I_{\{COMP1\}}(IN1, 1)$ is the resulting damage to *IN1*. We recall once again our assumption of *Information*

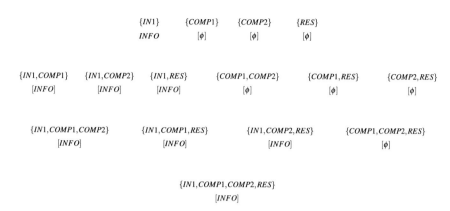

Fig. 9 The knowledge sets at time $t = 0$

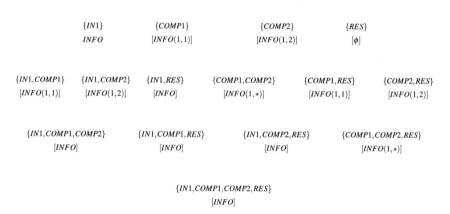

Fig. 10 The knowledge sets at time $t = 1$

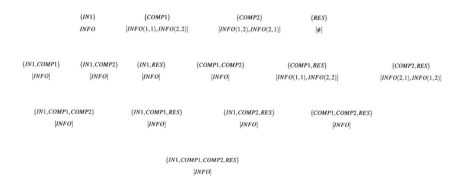

Fig. 11 The knowledge sets at time $t = 2$

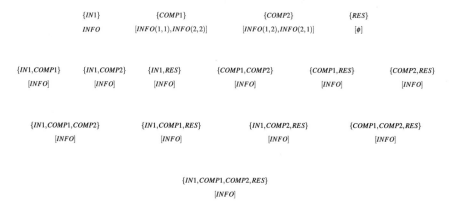

Fig. 12 The knowledge sets at time $t = 3$

completeness (Sect. 4): given a message delivery (A_s, A_d, m_{ts}), with $m_{ts} \leq INFO_s$, then $\beta_i(E_{sd}) = 0$. This property expresses an estimated possibility of zero that the sharing generation scheme can be broken. Therefore $IN1$ will attribute no risk to this stage, where no subset of actors not including itself has knowledge of both $INFO_1$ shares. Once again, we remark that a weaker version of the *information completeness* assumption can be adopted here to reflect the weakness of the naive share generation scheme, which could be easily broken by $COMP1$ via an educated guess. However, the threat space under consideration does not include autonomous guesses, and the original assumption is kept. At $t = 2$, unlike the previous example, no other singleton exists such that $K_X(1) = INFO_1$. However, this time other subsets for which $K_X(1) = INFO_1$ can be obtained, namely $\{COMP1, COMP2\}$.

Risk estimate at $t = 2$ by $IN1$ is therefore:

$$R(A_k, E_S) = R(IN1, E_{\{COMP1, COMP2\}})$$

that becomes:

$$R(A_k, E_S) = P_{\{COMP1\}}(1)I_{\{COMP1\}}(IN1, 1) \\ + P_{\{COMP1, COMP2\}}(2)I_{\{COMP1, COMP2\}}(IN1, 2) \tag{6}$$

$P_{\{COMP1, COMP2\}}$ is the possibility of the event where $\{COMP1, COMP2\}$ will actually share the information they have to reconstruct $INFO[1]$ multiplied by the damage $IN1$ would incur in, should the disclosure event actually happen. Note that, if this possibility is considered zero by default (e.g. the assessor is completely sure that $COMP1$ and $COMP2$ do not know of each other) then the risk at $t = 2$ is also 0.

7 Conclusions

The possibilistic risk analysis methodology presented in this chapter provides a fresh look at fully quantitative risk management on the cloud, enabling the comparison of cloud-based process models including different security mechanisms from the point of view of the changes in risk they imply. Our approach can be extended to cover most "cost versus risk" assessment activities. Also, the process model used in our methodology gracefully extends existing machine-readable specification of processes like the W3C candidate recommendation for choreographies WS-CDL (http://www.w3.org/2002/ws/chor/) and is supported by an open source software toolkit (http://sesar.di.unimi.it/cloudrisk) integrating a choreography editor.

Acknowledgements This work was partly supported by the European Commission within the PRACTICE project (contract n. FP7-609611) by the Italian MIUR project SecurityHorizons (c.n. 2010XSEMLC) and by the CMIRA2014/AcceuilPro (Subv. 14.004390) and COOPERA program of the Region Rhone-Alpes, France.

References

1. Winkler, V.: Cloud computing: risk assessment for the cloud. Technet Magazine, 01/2012
2. O'Hagan, A., Oakley, J.E.: Probability is perfect, but we can't elicit it perfectly. Reliab. Eng. Syst. Safety **85**(13), 239–248 (2004) (Alternative Representations of Epistemic Uncertainty)
3. Baudrit, C., Couso, I., Dubois, D.: Joint propagation of probability and possibility in risk analysis: towards a formal framework. Int. J. Approx. Reasoning **45**(1), 82–105 (2007)
4. Zadeh, A.L.: Fuzzy sets as a basis for a theory of possibility. Fuzzy Sets Syst. **1**, 3–28 (1978)
5. Dubois, D.: Fuzzy Sets and Systems: Theory and Applications, v.144. Academic press, New York (1980)
6. Dubois, D., Prade, H.: Default reasoning and possibility theory. Artif. Intell. **35**(2), 243–257 (1988)
7. De Cooman, G.:Possibility theory i: the measure-and integral-theoretic groundwork. Int. J. Gen. Syst. **25**(4), 291–323 (1997); Gert De Cooman. Possibility theory ii: Conditional possibility. International Journal Of General System, 25(4):325–351, 1997; Gert De Cooman. Possibility theory iii: Possibilistic independence. International Journal of General Systems, 25:353–372, 1997
8. Couso, I., Dubois, D., Sanchez, L.: Random Sets and Random Fuzzy Sets As Ill-Perceived Random Variables. Springer Publishing Company Incorporated, Heidelberg (2014)
9. Dubois, D., Prade, H.: Possibility theory and its applications: a retrospective and prospective view. In: Decision Theory and Multi-Agent Planning pp. 89–109. Springer, Heidelberg (2006)
10. Dubois, D., Prade, H.: Possibility theory. Scholarpedia **2**(10), 2074 (2007)
11. Heiser, J., Nicolett, M.: Assessing the security risks of cloud computing (2008)
12. Drissi, S., Houmani, H., Medromi, H.: Survey: risk assessment for cloud computing. Int. J. Adv. Comput. Sci. Appl. **4**, 143–148 (2013)
13. Fitó, J.O., Guitart, J.: Introducing risk management into cloud computing. Technical Report UPC-DAC-RR-2010-33, Technical University of Catalonia (2010)
14. Sangroya, A., Kumar, S., Dhok, J., Varma, V.: Towards analyzing data security risks in cloud computing environments. In: Information Systems, Technology and Management—International Conference ICISTM 2010, Proceedings, pp. 255–265 (2010)

15. Catteddu, D., Hogben, G.: Cloud computing: benefits, risks and recommendations for information security. Technical report, ENISA 2009 at www.enisa.europa.eu
16. Cloud Security Alliance: Security guidance for critical areas of focus in cloud computing v2.1, Technical Report 2009
17. NIST: Recommended security controls for federal information systems (2009)
18. ATOS: Risk analysis framework for a cloud specific environment. www.atos.net (2008)
19. The Open Group: Risk taxonomy. www.opengroup.org (2008)
20. Gadia, Sailesh: Cloud computing risk assessment: A case study. ISACA Journal 1, 1–6 (2012)
21. Information Systems Audit and Control Association: Cobit 5. http://www.isaca.org/Knowledge-Center/Research/ResearchDeliverables/Pages/Risk-Scenarios-Using-COBIT-5-for-Risk.aspx (2013)
22. Saripalli, P., Walters, B.: QUIRC: A quantitative impact and risk assessment framework for cloud security. In: IEEE 3rd International Conference on Cloud Computing (CLOUD), pp. 280–288 (2010)
23. Sendi, A.S., Cheriet, M.: Cloud computing: a risk assessment model. In: IEEE International Conference on Cloud Engineering (IC2E), pp. 147–152 (2014)
24. Khan, A.U., Oriol, M., Kiran, M., Jiang, M., Djemame, K.: Security risks and their management in cloud computing. In: IEEE 4th International Conference on Cloud Computing Technology and Science (CloudCom), pp. 121–128 (2012)
25. den Braber, F., Brndeland, G., Dahl, H.E.I., Engan, I., da Hogganvik, I., Lund, M.S., Solhaug, B., Stlen, K., Vraalsen, F.: The coras model-based method for security risk analysis. Technical report, SINTEF, 2006
26. Information risk analysis methodology, IRAM
27. Cavoukian, A.: Privacy risk management. Technical report, Information and Privacy Commissioner—Ontario - Canada, 2010
28. Kung, A., Crespo Garcia, A., Notario McDonnell, N., Kroener, I., Le Mtayer, D., Troncoso, C., Mara del Lamo, J., Martns, Y.S.: Pripare: A new vision on engineering privacy and security by design. Technical report, PRIPARE (2014)
29. Wright, D.: Should privacy impact assessments be mandatory? Commun. ACM 54(8), 121–131 (2011)
30. Garcia, P.A.A., Schirru, R., et al.: A fuzzy data envelopment analysis approach for FMEA. Prog. Nucl. Energy 46(3), 359–373 (2005)
31. Gargama, H., Chaturvedi, S.K.: Criticality assessment models for failure mode effects and criticality analysis using fuzzy logic. Reliab. IEEE Trans. 60(1), 102–110 (2011)
32. Yang, Z., Bonsall, S., Wang, J.: Fuzzy rule-based Bayesian reasoning approach for prioritization of failures in FMEA. Reliab. IEEE Trans. 57(3), 517–528 (2008)
33. Mohamed, S., McCowan, A.K.: Modelling project investment decisions under uncertainty using possibility theory. Int. J. Project Manage. 19(4), 231–241 (2001)
34. Lorterapong, P., Moselhi, O.: Project-network analysis using fuzzy sets theory. J. Constr. Eng. Manage. 122(4), 308–318 (1996)
35. Wong, K.C., So, A.T.P.: A fuzzy expert system for contract decision making. Constr. Manage. Econ. 13(2), 95–103 (1995)
36. Tam, C.M., Fung, I.: Assessing safety performance by fuzzy reasoning. Asia Pacific Build. Constr. Manage. J. 2(1), 6–13 (1996)
37. Karimi, I., Hüllermeier, E.: Risk assessment system of natural hazards: a new approach based on fuzzy probability. Fuzzy Sets Syst. 158(9), 987–999 (2007) (Selected papers from {IFSA} 2005 11th World Congress of International Fuzzy Systems Association)
38. Dubois, D., Prade, H.: Possibility theory in information fusion. Data fusion and perception. In: International Centre for Mechanical Sciences, vol. 431, pp. 53–76. Springer, Heidelberg (2001)
39. Dubois, D., Prade, H.: On the use of aggregation operations in information fusion processes. Fuzzy Sets Syst. 142(1), 143–161 (2004)
40. Dubois, D.: Representation, propagation, and decision issues in risk analysis under incomplete probabilistic information. Risk Anal. 30(3), 361–368 (2010)

41. Pedroni, N., Zio, E.: Empirical comparison of methods for the hierarchical propagation of hybrid uncertainty in risk assessment, in presence of dependence. Int. J. Uncertainty Fuzziness Know. Based Syst. **20**(04), 509–557 (2012)
42. Bilgiç, T., Türksen, I.B.: Measurement of membership functions: theoretical and empirical work. In: Fundamentals of fuzzy sets, pp. 195–227. Springer, Heidelberg (2000)
43. Zadeh, A.L.: Fuzzy sets. Inf. Control **8**(3), 338–353 (1965)
44. He, L., Xiao, J., Huang, H.-Z., Luo, Z.: System reliability modeling and analysis in the possibility context. In: IEEE International Conference on Quality, Reliability, Risk, Maintenance, and Safety Engineering (ICQR2MSE), pp. 361–367 (2012)
45. Huang, H.-Z., Tong, X., Zuo, M.J.: PosBist fault tree analysis of coherent systems. Reliab. Eng. Syst. Saf. **84**(2), 141–148 (2004)
46. He, L., Huang, H., Du, L., Zhang, X., Miao, Q.: A review of possibilistic approaches to reliability analysis and optimization in engineering design. In: Human-Computer Interaction. HCI Applications and Services, Lecture Notes in Computer Science, vol. 4553, pp. 1075–1084. Springer, Heidelberg (2007)
47. Onisawa, T.: An approach of system reliability analysis using failure possibility and success possibility. In: IV IEEE International Conference on Fuzzy Systems and II International Fuzzy Engineering Symposium, Proceedings of 1995 IEEE vol. 4, pp. 2069–2076 (1995)
48. Guyonnet, D., Bellenfant, G., Bouc, O.: Soft methods for treating uncertainties: applications in the field of environmental risks. In: Soft Methods for Handling Variability and Imprecision, Advances in Soft Computing, vol. 48, pp. 16–26. Springer, Heidelberg (2008)
49. Bortolan, G., Degani, R.: A review of some methods for ranking fuzzy subsets. Fuzzy Sets Syst. **15**(1), 1–19 (1985)
50. Dubois, D., Prade, H.: Ranking fuzzy numbers in the setting of possibility theory. Inf. Sci. **30**(3), 183–224 (1983)
51. Basu, S., Bultan, T.: Choreography conformance via synchronizability. In Proceedings of International Conference on World Wide Web, WWW 2011, pp. 795–804 (2011)
52. Bellare, M., Hoang, V.T., Rogaway, P.: Foundations of garbled circuits. In: The ACM Conference on Computer and Communications Security, CCS'12, Raleigh, NC, pp. 784–796 (2012)
53. Rabin, M.O.: How to exchange secrets with oblivious transfer. IACR Cryptology ePrint Archive **2005**, 187 (2005)
54. CISCO: Data leakage worldwide white paper: the high cost of insider threats (2011)
55. In *Networks and Groups*, Studies in Economic Design. (2003)
56. Anisetti, M., Bellandi, V., Damiani, E., Frati, F., Gianini, G., Jeon, G., Jeong, J.: Supply chain risk analysis: open source simulator. In Proceedings of V International Conference Signal Image Technology and Internet Based Systems, SITIS, pp. 443–450 (2009)
57. Anisetti, M., Damiani, E., Frati, F., Cimato, S., Gianini, G.: Using incentive schemes to alleviate supply chain risks. In: Proceedings of International Conference on Management of Emergent Digital Eco Systems, MEDES'10, pp. 221–228. ACM, New York, NY (2010)
58. Ceravolo, P., Damiani, E., Fasoli, D., Gianini, G.: Representing immaterial value in business model. In: Enterprise Distributed Object Computing Conference Workshops 2010, pp. 323–329
59. Damiani, E.: Risk-aware collaborative processes. In: International Conference on Enterprise Information Systems (ICEIS). ISBN 978-989-8111-88-3 (2009)
60. Damiani, E., Frati, F., Tchokpon, R.: The role of information sharing in supply chain management: the securescm approach. Int. J. Innov. Technol. Manage. **08**(03), 455–467 (2011)
61. Frati, F., Damiani, E., Ceravolo, P., Cimato, S., Fugazza, C., Gianini, G., Marrara, S., Scotti, O.: Hazards in full-disclosure supply chains. In: Conference on Advanced Information Technologies for Management (AITM). Publishing house of the Wroclaw University of Economics (2008)

62. Kerschbaum, F., Pibernik, R., Damiani, E., Gianini, G.: Toward value-based control of knowledge sharing in networked services design. Prace Naukowe Uniwersytetu Ekonomicznego we Wrocllawiu **85**, 51–65 (2009)
63. Chan, S.Y.: An alternative approach to the modeling of probability distributions. Risk Anal. **13**(1), 97–102 (1993)
64. van Dorp, J.R., Rambaud, S.C., Pérez, J.G., Pleguezuelo, R.H.: An elicitation procedure for the generalized trapezoidal distribution with a uniform central stage. Decis. Anal. **4**(3), 156–166 (2007)
65. MacDonell, S.G., Gray, A.R., Calvert, J.M.: FULSOME: A fuzzy logic modeling tool for software metricians. In: IEEE International Conference of the North American Fuzzy Information Processing Society, NAFIPS 1999, pp. 263–267 (1999)
66. Howard, R.A.: Information value theory. IEEE Trans. Sys. Science Cybern. **2**(1), 22–26 (1966)
67. Hoomans, T., Seidenfeld, J., Basu, A., Meltzer, D.: Systematizing the use of value of information analysis in prioritizing systematic reviews. Technical Report 12-EHC109-EF, Agency for Healthcare Research and Quality (2012)
68. Cheng, P.C., Rohatgi, P., Keser, C., Karger, P.A., Wagner, G.M., Reninger, A.S.: Fuzzy multi-level security: an experiment on quantified risk-adaptive access control. In: IEEE Symposium on Security and Privacy, SP'07, 2007, pp. 222–230
69. Chowdhury, S., Champagne, P., McLellan, P.J.: Uncertainty characterization approaches for risk assessment of {DBPs} in drinking water: a review. J. Environ. Manage. **90**(5), 1680–1691 (2009)
70. Gupta, A., Maranas, C.D.: Managing demand uncertainty in supply chain planning. Comput. Chem. Eng. **27**(89), 1219–1227 (2003) (2nd Pan American Workshop in Process Systems Engineering)
71. Hanratty, T., Hammell II, R.J., Heilman, E.: A fuzzy-based approach to the value of information in complex military environments. In Scalable Uncertainty Management, Lecture Notes in Computer Science, vol. 6929, pp. 539–546. Springer, Heidelberg (2011)
72. Tanaka, H., Ichihashi, H., Asai, K.: A value of information in FLP problems via sensitivity analysis. Fuzzy Sets Syst. **18**(2), 119–129 (1986)
73. Scheer, A.-W., Nüttgens, M.: ARIS architecture and reference models for business process management. In: Business Process Management, Models, Techniques, and Empirical Studies, pp. 376–389 (2000)
74. Bogdanov, D., Kamm, L., Laur, S., Pruulmann-Vengerfeldt, P.: Secure multi-party data analysis: end-user validation and practical experiments. Cryptology ePrint Archive, Report 2013/826 (2013)
75. Buckley, I., Fernández, E.B., Anisetti, M., Ardagna, C.A., Sadjadi, S.M., Damiani, E.: Towards pattern-based reliability certification of services. In: On the Move to Meaningful Internet Systems Proceedings, Part II, pp. 560–576 (2011)
76. Damiani, E., Ardagna, C.A., EI Ioini, N.: Open Source Systems Security Certification. Springer, Heidelberg (2009). ISBN 978-0-387-77323-0
77. Kolesnikov, V.: Gate evaluation secret sharing and secure one-round two-party computation. In: Advances in Cryptology—ASIACRYPT 2005, 11th International Conference on the Theory and Application of Cryptology and Information Security, Proceedings, pp. 136–155 (2005)
78. Malkhi, D., Nisan, N., Pinkas, B., Sella, Y.: Fairplay—secure two-party computation system. In: Proceedings of the 13th USENIX Security Symposium, August 9–13, 2004, San Diego, CA, USA, pp. 287–302 (2004)
79. Schneider, T., Zohner, M.: GMW vs. Yao? Efficient secure two-party computation with low depth circuits. In: Financial Cryptography and Data Security—17th International Conference FC 2013, Okinawa, Japan, April 1–5, 2013, Revised Selected Papers, pp. 275–292 (2013)

Author Biographies

Valerio Bellandi received the M.S. and Ph.D. degrees in computer science at the Università degli Studi di Milano (Italy) in 2004 and 2009, respectively. He is currently a Post-Doctoral Fellow in the Department of Computer Science of the Università degli Studi di Milano and he is a member of the staff of the SESAR LAB. His main research interests and in the Area of Collaborative Environment, Mobile Networks and Business Processes. He has published several papers in journals and conferences and has served in the program committee of several international conferences.

Stelvio Cimato is an Assistant Professor with the Dipartimento di Informatica of the Università degli Studi di Milano (Italy). He got the Ph.D. in Computer Science at the Università degli Studi di Bologna (Italy) in 1999. His main research interests are in the area of cryptography, network security, and Web applications. He has published several papers in the field and is active in the community, serving as member of the program committee of several international conferences in the area of cryptography and data security.

Ernesto Damiani is a full professor at Università degli Studi di Milano (Italy), the director of the University's Ph.D. program in computer science and the coordinator of SESAR Research Lab (http://sesar.di.unimi.it). He has held visiting positions at a number of international institutions, including George Mason University in Virginia, US, La Trobe University in Melbourne, Australia, Tokyo Denki University, Japan and INSA-Lyon, France. Ernesto Damiani's research interests include cloud assurance, Web services and business process security and Big Data processing. He has served as Chair of many conferences, including the IEEE International Conference on Web Services (ICWS), the IEEE Conference on Digital Ecosystems series (IEEEDEST) and the IFIP Working Conference on Open Source Systems (OSS). Ernesto Damiani has published more than 300 papers, books and international patents. He is a senior member of the IEEE and a Distinguished Scientist of the ACM. He is the author of "Open Source Security Certification", Springer, 2009.

Gabriele Gianini has received a Ph.D. in Physics 1996 from the Università degli Studi di Pavia (Italy). Since 2005 is Assistant Professor at the Department of Computer Science of the Università degli Studi di Milano (Italy), where he is a lecturer of Probability and Statistics. He has held visiting positions at a number of international institutions, including INSA de Lyon (France), University of Passau (Germany), CERN (Switzerland), Fermilab (US), CBPF (Brasil). He has been Adjoint professor at the Free University of Bolzano (Italy) from 2005 to 2012. Among his research interests: game theoretic applications to networking and to security, quantitative modeling of processes, statistical and soft computing techniques. He has been conducting research activities within several FP7 projects. From 1992 to present has co-authored over 150 papers published in internationally refereed journals.